Mathematics
A PRACTICAL GUIDE STUDENT BOOK

Talei Kunkel
Palmira Mariz Seiler
Marlene Torres-Skoumal
David Weber

International Baccalaureate
Baccalauréat International
Bachillerato Internacional

Mathematics: A practical guide (Student book)

Published on behalf of the International Baccalaureate Organization, a not-for-profit educational foundation of 15 Route des Morillons, 1218 Le Grand-Saconnex, Geneva, Switzerland by the

Published by International Baccalaureate Organization (UK) Ltd, Peterson House, Malthouse Avenue, Cardiff Gate, Cardiff, Wales CF23 8GL United Kingdom, represented by IB Publishing Ltd, Churchillplein 6, The Hague, 2517JW The Netherlands
Website: www.ibo.org

The International Baccalaureate Organization (known as the IB) offers four high-quality and challenging educational programmes for a worldwide community of schools, aiming to create a better, more peaceful world. This publication is one of a range of materials produced to support these programmes.

IB merchandise and publications can be purchased through the IB store at http://store.ibo.org. General ordering queries should be directed to the Sales and Marketing Department at sales@ibo.org.

British Library Cataloguing in Publication Data

A catalogue record for this book is available from the British Library

ISBN: 978-0-9927035-0-9
MYP338
Typeset by Q2A Media Services Pvt Ltd
Printed and bound in Dubai.

Acknowledgments

We are grateful for permission to reprint copyright material and other content:

p6 blocks: ©iStockphoto, tools: ©iStockphoto, plans: ©iStockphoto, house: ©iStockphoto; p10 treehouse: ©iStockphoto; p25 towers of Hanoi: http://commons.wikimedia.org/wiki/Category:Tower_of_Hanoi; p27 paper: ©iStockphoto; p35 bull and bear: http://commons.wikimedia.org/wiki/File:Bulle_und_B%C3%A4r_Frankfurt.jpg; p49 posts: ©iStockphoto; p58 Königsberg: http://commons.wikimedia.org/wiki/File:Konigsberg_bridges.png; p71 Euclid: http://commons.wikimedia.org/wiki/File:EuclidStatueOxford.jpg; p72 solar system: ID 21096100 © Jut | Dreamstime.com; p79 space shuttle: http://commons.wikimedia.org/wiki/File:Space_Shuttle_Columbia_launching.jpg; p83 tree: ©iStockphoto; p86 Austrian balloons: ©Family Koller; p87 single balloon: ©Family Koller; p102 Shanghai skyline: ©iStockphoto; p108 sprinklers: ©iStockphoto; p110 urban planning: ©iStockphoto; p113 Lorenz attractor: © Paul Bourke; p123 Mona Lisa: http://commons.wikimedia.org/wiki/File:Mona_Lisa,_by_Leonardo_da_Vinci,_from_C2RMF_retouched.jpg; p124 Parthenon: http://commons.wikimedia.org/wiki/Parthenon#mediaviewer/File:2006_01_21_Ath%C3%A8nes_Parth%C3%A9non.JPG; p136 Aristotle: ©iStockphoto; p139 cup: ©iStockphoto; p161-2 Arctic sea ice graph: ©National Snow and Ice Data Center, University of Colorado, Boulder; p166 art gallery: ©iStockphoto; p166 painting ©iStockphoto; p168 Napier's bones: http://commons.wikimedia.org/wiki/File:%C3%81bacos_neperianos_[M.A.N._Madrid]_03.jpg?fastcci_from=5378149; p173 tree trunk: ©iStockphoto; p180 mobile: ©iStockphoto; p181 UFO: ©iStockphoto; p183 Porto fountain: Image courtesy of Adam Fuller, http://www.flickr.com/photos/adamfrunski; p184 The Madelbrot set: ©iStockphoto; p201 blood donation: http://commons.wikimedia.org/wiki/File:Blood_Donation_at_ITESM_CCM.jpg.

IB learner profile

The aim of all IB programmes is to develop internationally minded people who, recognizing their common humanity and shared guardianship of the planet, help to create a better and more peaceful world.

As IB learners we strive to be:

INQUIRERS
We nurture our curiosity, developing skills for inquiry and research. We know how to learn independently and with others. We learn with enthusiasm and sustain our love of learning throughout life.

KNOWLEDGEABLE
We develop and use conceptual understanding, exploring knowledge across a range of disciplines. We engage with issues and ideas that have local and global significance.

THINKERS
We use critical and creative thinking skills to analyse and take responsible action on complex problems. We exercise initiative in making reasoned, ethical decisions.

COMMUNICATORS
We express ourselves confidently and creatively in more than one language and in many ways. We collaborate effectively, listening carefully to the perspectives of other individuals and groups.

PRINCIPLED
We act with integrity and honesty, with a strong sense of fairness and justice, and with respect for the dignity and rights of people everywhere. We take responsibility for our actions and their consequences.

OPEN-MINDED
We critically appreciate our own cultures and personal histories, as well as the values and traditions of others. We seek and evaluate a range of points of view, and we are willing to grow from the experience.

CARING
We show empathy, compassion and respect. We have a commitment to service, and we act to make a positive difference in the lives of others and in the world around us.

RISK-TAKERS
We approach uncertainty with forethought and determination; we work independently and cooperatively to explore new ideas and innovative strategies. We are resourceful and resilient in the face of challenges and change.

BALANCED
We understand the importance of balancing different aspects of our lives—intellectual, physical, and emotional—to achieve well-being for ourselves and others. We recognize our interdependence with other people and with the world in which we live.

REFLECTIVE
We thoughtfully consider the world and our own ideas and experience. We work to understand our strengths and weaknesses in order to support our learning and personal development.

The IB learner profile represents 10 attributes valued by IB World Schools. We believe these attributes, and others like them, can help individuals and groups become responsible members of local, national and global communities.

International Baccalaureate®
Baccalauréat International
Bachillerato Internacional

Contents

How to use this book   vi

1. Introduction to IB Skills   1

2. Introducing key concept 1: form   8

3. Introducing key concept 2: logic   11

4. Introducing key concept 3: relationships   17

5. Change   21

       Topic 1: Finite patterns of change   23

       Topic 2: Infinite patterns of change   29

       Topic 3: Financial mathematics   32

6. Equivalence   37

       Topic 1: Equivalence, equality and congruence   38

       Topic 2: Equivalent expressions and equivalent equations   41

       Topic 3: Equivalent methods and forms   44

7. Generalization   51

       Topic 1: Number investigations   54

       Topic 2: Diagrams, terminology and notation   56

       Topic 3: Transformations and graphs   65

8. Justification   70

       Topic 1: Formal justifications in mathematics   71

       Topic 2: Empirical justifications in mathematics   77

       Topic 3: Empirical justifications using algebraic methods   79

9. Measurement   82

       Topic 1: Making measurements   84

       Topic 2: Related measurements   88

       Topic 3: Using measurements to determine inaccessible measures   92

10. Model   97

       Topic 1: Linear and quadratic functions   99

       Topic 2: Regression models   102

       Topic 3: Scale models   108

11. Pattern   112

       Topic 1: Famous number patterns   114

Topic 2: Algebraic patterns 124

Topic 3: Applying number patterns 126

12. Quantity 135

Topic 1: Volumes, areas and perimeters 138

Topic 2: Trigonometry 140

Topic 3: Age problems 147

13. Representation 151

Topic 1: Points, lines and parabolas 152

Topic 2: Probability trees 158

Topic 3: Misrepresentation 160

14. Simplification 164

Topic 1: Simplifying algebraic and numeric expressions 166

Topic 2: Simplifying through formulas 169

Topic 3: Simplifying a problem 172

15. Space 178

Topic 1: Special points and lines in 2D shapes 180

Topic 2: Mathematics and art 182

Topic 3: Volume and surface area of 3D shapes 187

16. System 189

Topic 1: The real number system 191

Topic 2: Geometric systems 195

Topic 3: Probability systems 197

How to use this book

As well as introducing you to the 3 key concepts and 12 of the related concepts in the Middle Years Programme (MYP) mathematics course, this book will also help you practise all the skills you need to reach the highest level of the MYP assessment criteria.

This book has been divided into chapters on key and related concepts. Throughout the book you will find features that will help you link your learning to the core elements of the MYP.

On the first page of each of the related concept chapters you will find:

- the topics you will be focusing on
- the inquiry questions you will be considering
- a checklist of skills you will practice
- a glossary of any difficult terms
- a list of the command terms you will come across.

You will also see a list of other concepts that relate to the chapter. You should keep these in mind as you work.

Every chapter begins with a quote from a mathematician related to the concept being developed. Each related concept chapter is divided into three topics that help you explore the concept through a variety of activities. Some activities can be done individually while others may be done with a partner or in a group. At the end of each topic is a reflection where you are given the opportunity to think about what you have learned and how it may relate to what you already know.

Here are some of the other features that you will come across in the book.

GLOBAL CONTEXTS

For many of the activities you will see an indication of a global context that is the focus of that activity. Global contexts help organize inquiry into six different areas.

- Identities and relationships
- Orientation in space and time
- Personal and cultural expression
- Scientific and technical innovation
- Globalization and sustainability
- Fairness and development

These global contexts indicate how the activity is relevant to your life and the real world.

ATL SKILLS

Alongside global contexts, each topic and activity includes an ATL skills focus. Usually, there are only one or two skills identified as the focus for an activity or topic. Of course, you will be using and developing other skills, but there is an emphasis on the particular skills in these boxes.

TIP

Throughout the chapters you will see additional information to help your understanding of a topic or activity.

QUICK THINK

These boxes provide questions to challenge your thinking. Your teacher may use them for a class discussion.

INTERDISCIPLINARY LINKS

As an MYP student you are encouraged to use skills and knowledge from different subject areas in your learning. Look out for these boxes which provide links to other subject groups.

CHAPTER LINKS

These boxes direct you to other chapters that relate to a topic or activity.

WEB LINKS

These boxes include websites and search terms for further reading and exploration.

INTERNATIONAL MATHEMATICS

These boxes identify mathematical terms that differ in various parts of the world.

MATHS THROUGH HISTORY

These boxes introduce historical context for mathematical concepts included in a topic or activity.

Introduction to IB skills

Welcome to *Mathematics: A Practical Guide* for MYP 4/5. The purpose of this book is to help IB students develop and apply mathematical skills through the use of student-centred activities and authentic, real-life tasks. Rather than being aimed at one particular course, this book includes a variety of content from all four branches of the Middle Years Programme (MYP) mathematics framework: number, algebra, geometry and trigonometry, and statistics and probability.

Take a look at the table of contents. You will notice that the chapters aren't like those in a typical textbook, where each chapter focuses on one topic, for example, fractions. These chapters are organized by concept. In the MYP, there are three key concepts, which you will briefly explore in chapters 2–4.

Key concepts

In the MYP, each subject area has key concepts that are used as a framework for knowledge within that subject area. They are powerful ideas that you will explore, through different topics, to try to understand the world around you. In MYP mathematics, there are three key concepts that are the basis for study: form, logic and relationships.

Form refers to the underlying structure and shape of something and how it is distinguished by its properties. For example, different graphs in statistics are different forms of data representation, each with its own characteristics.

Logic is a process used in reaching conclusions about numbers, shapes and variables. For example, finding the sizes of angles in a diagram requires a logical process.

Relationships are the connections between quantities, properties or concepts that can be expressed as models, rules or statements. For example, a relationship enables you to graph data and model it with a linear function.

Notice how these are quite general. If you heard that a teacher was teaching a unit on relationships, you might not immediately assume it was a mathematics class. It could just as easily be science or language and literature or even individuals and societies.

Related concepts

There are also 12 related concepts, which are the central themes for the chapters in this book. These concepts appear to be much more related to mathematics.

Related concepts in mathematics		
Change	Equivalence	Generalization
Justification	Measurement	Model
Pattern	Quantity	Representation
Simplification	Space	System

Related concepts in mathematics

The teacher whose current unit is about relationships could be from a variety of subject areas. If the related concepts are "character" and "setting", then that teacher is likely to be a language and literature specialist. If, however, the related concepts are "quantity" and "generalization", then it is a reasonably good assumption that this is a mathematics teacher.

How will your teachers use these concepts in their teaching? While your units of study in class will probably still be organized by mathematical topic, your teachers will select a key concept and several related concepts for each unit. How do they know which ones to choose? There really is no right answer, since the choice depends on how your teachers want you to understand that content. Looking at a topic as a logical process instead of a series of relationships or as a structure changes the focus and how it is taught. Let's try some examples.

 Activity 1 **Choosing concepts**

For each unit listed below, pick the key concept that you think is the most appropriate. Then add a few related ones that fit well with it. Here is an example, to get you started.

EXAMPLE

UNIT: Solving one- and two-step equations

a) Is solving equations about a process? Is it more focused on the relationships between the simplified equations you write as you solve them? Is it all about the structure and properties of the equations? Answering these questions will help you arrive at the key concept that will be the focus of the unit. Choose only one.

b) Explain which related concepts from the above table are also explored in a unit on solving equations.

Work in a small group. Select a key concept and a few related concepts for each unit.

UNIT: Decimals, ratios and percentages

UNIT: Perimeter and area of two-dimensional shapes

UNIT: Order of operations

UNIT: Mean, median and mode

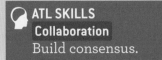
ATL SKILLS
Collaboration
Build consensus.

As an MYP mathematics student you can experience the joy of discovery and success in mathematics and then apply your new knowledge to the world around you. You can better understand the mathematics you are learning because you have, mainly, formulated the concepts yourself rather than just hearing about them. You are more willing to persist in problem-solving because you are used to meeting new content or new contexts that challenge you to work through any difficulties. At the same time, you are developing ways of learning that would be almost impossible if you were a passive participant in a classroom. In an MYP mathematics class, you aren't just learning mathematics. You are acquiring skills and techniques that will serve you well in school and throughout life while, at the same time, learning how to learn.

Learning skills

Now that you understand what the "conceptual" in conceptual learning means, take a closer look at the word "learning". You know that you have been learning all your life. First, you began in settings such as your home and neighbourhood. Then, your learning became more formal as you started school. Learning in the MYP is primarily inquiry-based learning. Your learning will continually cycle through three different phases.

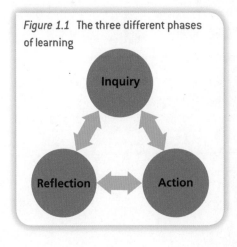

Figure 1.1 The three different phases of learning

Inquiry

Ask questions! It's the only way you are going to find out exactly what you want to know. Think about what you already know and what you want to know. Your curiosity is one of your best assets as a student.

Action

An important part of conceptual learning is action. Action in the MYP might involve learning by doing, service learning, educating yourself and educating others. Sometimes you may choose to act, based on newly acquired knowledge and understandings. Remember to be principled in your actions and make responsible choices.

Reflection

As a learner, you will become increasingly aware of the way in which you use evidence, practise skills and make conclusions. Reflection in your learning helps you to look at the facts from a different perspective, to ask new questions and to reconsider your own conclusions. You may then decide to lead your inquiry in a different direction.

Inquiry learning can be frustrating. There is not always a right answer; sometimes conclusions may be uncomfortable or may conflict with what you want to believe. You will realize that there are no endpoints in learning. As an MYP student, learning through inquiry, action and reflection is central to your education; it forms the foundation of acquiring knowledge and conceptual understanding.

Conceptual learning is:	Conceptual learning is not:
learning through inquiry	learning only through memorization
taking action to understand the world around you	trying to find the "right" answer
using knowledge to understand big ideas	passively accepting everything you read, hear or see.
making connections through concepts across different subjects.	

The characteristics of conceptual learning

Knowledge

A concept such as systems or relationships isn't something you can touch, but you can certainly use specific examples from different subject areas to explain it to another person. This is where your knowledge of facts is essential. Without the support of specific knowledge, facts and examples, it is very difficult to understand and explain key concepts and related concepts. In the MYP, your teachers can choose which facts and examples they will use to help develop your understanding of key concepts.

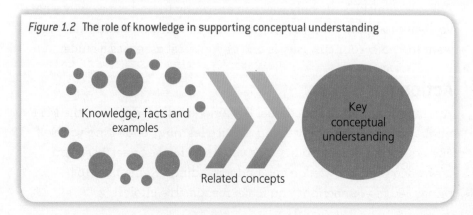

Figure 1.2 **The role of knowledge in supporting conceptual understanding**

Knowledge, facts and examples

Related concepts

Key conceptual understanding

The use of knowledge, facts and examples will be different in every MYP classroom but it will always lead you to an understanding of the key and related concepts in the subject group of mathematics.

Global contexts

Now that you know what the key and related concepts are, you can focus a little more on the knowledge, facts and examples that will help you understand, explain and analyse them. The MYP calls this part of the curriculum "global contexts". The global context is the setting or background for studying the key and related concepts. It is easy to think that the global context is the choice of topic in your course of study. There are six global contexts:

- identities and relationships
- orientation in space and time
- personal and cultural expression
- scientific and technical innovation
- globalization and sustainability
- fairness and development.

🌐 GLOBAL CONTEXTS

The choice of global context is influenced in several different ways.

Scale—study of a concept on an individual, local or global level.

Relevance—your education needs to be relevant for you and the world you live in, and this will influence the choice of context.

International mindedness—IB programmes aim to develop internationally minded students and this is supported through using a variety of contexts to understand concepts.

Do you, as a student, have influence over which global context is chosen? Of course you do! It's the reason why MYP mathematics courses look different all around the world. The contexts that are relevant for you may not be relevant for a student studying in another country or even in another school in your own country. What all MYP mathematics courses have in common is the goal of deepening your understanding of the mathematical key concepts.

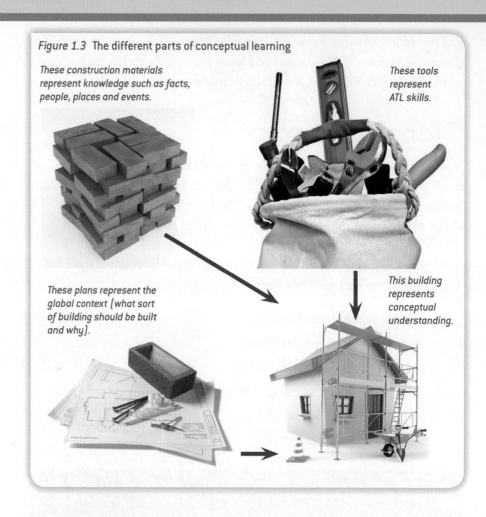

Figure 1.3 The different parts of conceptual learning

These construction materials represent knowledge such as facts, people, places and events.

These tools represent ATL skills.

These plans represent the global context (what sort of building should be built and why).

This building represents conceptual understanding.

Approaches to learning (ATL) skills

As a learner, you are developing a range of skills to help you learn and process significant amounts of knowledge and understanding. Some skills are very specific to particular subjects while others are those that you use every day in every class, and will ultimately use for life! The skills that you learn through the MYP allow you to take responsibility for your own learning. There are five groups of MYP skills.

ATL SKILL	Communication
Thinking	Self-management
Social	Research

Depending on the subject, you might focus more on one or two areas than on others. As you move through the MYP and mature as a student, the focus will also move through different skills—from being taught, to practising—to consolidate your skill ability. Read through the outline of ATL skills, taking some time to consider where and when you have learned, practised or mastered different skills. Also, think about which skills you still need to learn, practise or master.

Thinking skills	Critical thinking—the skill of analysing and evaluating issues and ideas.
	Creative thinking—the skills of exercising initiative to consider challenges and ideas in new and adapted ways.
	Transfer—the skill of learning by making new connections and applying skills, knowledge and understanding to new situations.
Social skills	Collaboration—the skill of working cooperatively with others.
Communication skills	Interaction—the skill of effectively exchanging thoughts, messages and information.
	Literacy—the skills of reading, writing and using language to communicate information appropriately and write in a range of contexts.
Self-management skills	Organization—the skill of effectively using time, resources and information.
	Affective skills—the skills of managing our emotions through cultivating a focused mind.
	Reflection—the skill of considering and reconsidering what is learned and experienced in order to support personal development through metacognition.
Research skills	Information literacy and Media literacy—the skills of interpreting and making informed judgments as users of information and media, as well as being a skillful creator and producer of information and media messages.

Approaches to learning (ATL) skills

It would be impossible to focus on all these areas in just your MYP mathematics course in years 4 and 5, so specific skills to learn, practise and master have been selected for inclusion in this book.

Summary

Look back to Figure 1.3 on page 6. Remember that conceptual learning happens when you use the inquiry cycle, develop your ATL skills and increase subject knowledge. These three factors work together to develop detailed understanding of the three key concepts in mathematics: form, relationships and systems. While the content of mathematics courses will look different in every MYP classroom, there is always the same focus on conceptual learning to construct a deeper understanding of the big ideas in life and the world around you.

References

Erickson, HL. 2008. *Stirring the Head, Heart and Soul: Refining Curriculum, Instruction, and Concept-Based Learning*. Thousand Oaks, California, USA. Corwin Press.

Introducing key concept 1: form

INQUIRY QUESTIONS	■ **What is the advantage of expressing something in a particular way?** ■ **Do different forms lead to different solutions?**
SKILLS	**ATL** ✓ Organize and depict information logically. ✓ Consider content. **Algebra** ✓ Rewrite quadratic functions in various forms. ✓ Solve quadratic equations in various ways. ✓ Solve a real-world problem.

GLOSSARY

Quadratic a polynomial expression of degree two.

COMMAND TERMS

Explain give a detailed account including reasons or causes.

Solve obtain the answer(s) using algebraic and/or numerical and/or graphical methods.

State give a specific name, value or other brief answer without explanation or calculation.

Introduction

The key concept of form in MYP mathematics refers to the understanding that the underlying structure and shape of an entity is distinguished by its properties. Exploring the key concept of form helps answer questions such as:

- How can we express mathematical relationships?
- What is the advantage of expressing something in a particular way?

Although you may not have realized it, you were studying form when you learned about fractions, decimals and percentages, which are three different ways of representing the same quantity.

As you learn more algebra, you will begin to see different forms for the equation of a line and different forms of representing a system of equations. Inevitably, the study of form will lead you to identifying different ways of representing a variety of functions, one of which is the focus of this chapter. If you believe the words of Einstein, constructing these different forms will then allow you to find them elsewhere, something that will also come to life in this chapter.

The human mind has first to construct forms, independently, before we can find them in things.

Albert Einstein

Activity 1   Forms of a quadratic function

A **quadratic** function is one that can be written in the form
$y = ax^2 + bx + c, a \neq 0$.

a) State the three forms of a quadratic function. Give an example of each.

b) Rewrite each of the given functions in the other forms. Show your work.

   i) $y = 2x^2 - 6x + 11$

   ii) $y = -3(x+1)^2 + 4$

   iii) $y = 3(x+1)(x-5)$

   iv) $y = -4x^2 + 16$

WEB LINKS
For two of the forms, go to http://www.mathopenref.com and search for "quadratic functions".

CHAPTER LINKS
See Chapter 6, Equivalence for more information about different forms of quadratic functions.

ATL SKILLS
Communication
Organize and depict information logically.

Activity 2   Solving quadratic equations

Problems based on quadratic functions usually require you to solve quadratic equations. You can use each of the forms you identified in activity 1 to solve a quadratic equation.

a) State the three methods of solving quadratic equations. Explain how they relate to the three forms of the quadratic function.

b) Construct a flow chart to represent the problem-solving process for solving a quadratic equation. Describe the method you use. Explain how you know when to use it. Make sure that you include all three forms in your flow chart.

ATL SKILLS
Communication
Organize and depict information logically.

Reflection

Identify functions other than quadratic functions that can be represented in a variety of forms. Describe some advantages and disadvantages of each form.

Grandfather is planning to dismantle an old treehouse, but before he does, he decides to keep a framed photo of it. He will enlarge the picture to 200 mm × 150 mm and use some wood from the treehouse to frame his photo.

A treehouse

After dismantling the treehouse, he decides to use a piece of wood that measures 255 mm × 125 mm. He wants to use the whole piece of wood for the frame. How wide does the frame need to be? The frame should be the same width, all the way around the picture, and no pieces should overlap. He may have to cut the piece of wood into more than four pieces before he assembles it to make the frame.

a) Sketch diagrams to show the information you are given. Add clear labels and define your variable(s).

b) **Explain** how you could use a quadratic equation to find the solution of this problem.

c) **Solve** the equation in two different ways.

d) **State** which form of the quadratic equation was the easiest to use. **Explain**.

e) Use a diagram to show how to cut the wood to make the frame.

⊂⊃ INTERDISCIPLINARY LINKS
Sciences

Analysing the motion of objects in physics often involves solving quadratic equations. Some of those equations can be factorized while others require the use of the quadratic formula.

⊕ GLOBAL CONTEXTS
Personal and cultural expression

◔ ATL SKILLS
Self-management
Consider content.

Reflection

Describe another way to solve the problem that doesn't require the use of quadratic equations.

Summary

Sometimes, the key concept of form can be very subtle. Thinking about form helps you to explore the reasons why you choose to represent quantities, relationships or information in certain ways. It encourages you to think about the advantages of one particular representation over another. It also lets you understand the structure of your chosen representation. Without referring to form, it is difficult to communicate mathematically or to make comparisons. For these reasons the concept of form is often at the core of any mathematics you will study.

INQUIRY QUESTIONS	■ How can logic be used to solve problems?
	■ Is logic ever wrong?
SKILLS	**ATL**
	✓ Demonstrate persistence and perseverance.
	✓ Listen actively to other perspectives and ideas.
	✓ Consider personal learning strategies.
	Number
	✓ Use logic to solve number puzzles.
	Geometry
	✓ Find the size of an unknown angle.

GLOSSARY

Calcudoku a mathematical puzzle in which you follow specific rules to fill in missing numbers in a grid.

COMMAND TERMS

Describe give a detailed account or picture of a situation, event, pattern or process.

Explain give a detailed account including reasons or causes.

Find obtain an answer showing relevant stages in the working.

Justify give valid reasons or evidence to support an answer or conclusion.

Introducing logic

In MYP mathematics, logic is a system of reasoning that helps you make decisions. You can use it to explain and justify the conclusions that you make. You have already used logic in deciding whether triangles were similar or congruent, and in classifying numbers as integers, rational, real and so on. You have used it as you moved across the different branches of mathematics (number, algebra, geometry and trigonometry, and statistics and probability).

In this book you will use logic as you learn new mathematical concepts such as patterns in expressions, numbers and shapes. It will help you to plan strategies and draw conclusions that are consistent with what you observe and calculate.

The more you practise using logic, the better you will become at it. Puzzles are fun and solving them is very rewarding – as well as being an excellent way of developing logical and mathematical reasoning.

Calcudoku puzzles are grid-based numerical puzzles that use the four basic mathematical operations (+, −, ×, ÷). The size of the grid

The rules of logic are to mathematics what those of structure are to architecture.

Bertrand Russell

ranges from 3 by 3 to 9 by 9, and they use different combinations of the operations (+, −, ×, ÷). They come in varying difficulty levels to ensure the puzzles can be enjoyed by people of all ages and abilities.

> ### 🏛 MATHS THROUGH HISTORY
> Calcudoku is the generic name of a puzzle that was developed in 2004 by a Japanese mathematics teacher, Tetsuya Miyamoto. His goal was to improve his students' mathematical reasoning and logic skills, as well as their concentration and perseverance. It quickly spread outside the classroom and is now enjoyed worldwide. Many newspapers and magazines, and countless websites, include similar puzzle challenges. In Japan, Calcudoku puzzles are called KenKen. "Ken" means "Wisdom" so KenKen is "Wisdom squared". In double Calcudokus, two puzzles are joined together, usually in one of the corners, so that some of the numbers are in both parts of the puzzle.

 Activity 1   Solving Calcudoku puzzles

Here are the basic rules and instructions for Calcudoku puzzles.

The numbers you can use in a puzzle depend on the size of the grid. For a 3 by 3 grid you can only use 1, 2 and 3. For a 4 by 4 grid, you can only use 1, 2, 3 and 4 and so on.

The heavily outlined groups of squares in each grid are called cages. In the upper-left corner of each cage, there is a target number and a mathematical operation (+, −, ×, ÷) that you must use to achieve the target number. You must combine the numbers in each cage to make the target number. The order of the numbers is not important. Suppose you have a 4 by 4 grid, the target number is 3 and the operation is subtraction, with a cage size of two squares. Then you can place the 4 and the 1 in any order you like within the cage.

The golden rule is that you can never repeat a number in any row or column. Remember that the solution must be unique.

Work individually or in pairs (to be decided by your teacher). Develop a strategy and a set of tips to help solve Calcudoku puzzles. As you are working through the easier ones, note the steps you complete first, in order to make the puzzle easier. Think about how you will record all of the possible combinations so you can analyse how to fit the numbers in a certain cage.

Look at some basic examples first. Try to complete each puzzle. Remember that you cannot repeat a number in any column or any row.

Puzzle 1 (left):

3 +	1 −	3
		2 ÷
3 ÷		

Puzzle 2 (right):

6 ×			4
12 ×	1 −		5 +
	2 ÷		
		2 −	

Now try a puzzle that is a little more challenging. As you fill in the squares, list the possible number combinations for each cage and look for patterns that can help you.

8 ×	1 −	4 −	2 −	
			10 +	
	1 −			2
2 −		2 ÷		2 −
10 ×			4	

REFLECTION

Work in a group of four. Together, plan a set of strategies to help solve these puzzles. Write them down, so that you can share them with the class. Are your strategies logical? **Justify** your answer.

GLOBAL CONTEXTS
Scientific and technical innovation

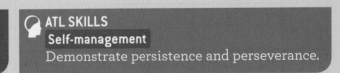

ATL SKILLS
Self-management
Demonstrate persistence and perseverance.

With the whole class, discuss and develop a master list of tips/rules to help you solve Calcudoku puzzles. Your teacher will give you an example. Work through it together.

STEP 1   Now use the strategies and tips you have developed to solve these puzzles. Are your strategies logical? **Justify** your answer.

Puzzle 1 (left):

1 −		3 +		1 −	6 ×
3 ÷	11 +		24 ×		
	8 +			80 ×	3 −
2 −					
	5 −		10 +		
10 ×		3 ÷		2 −	

Puzzle 2 (right):

24 ×		13 +		2 −	8 +
5 −		2 ÷			
120 ×	1 −		4 +	2 ÷	
					13 +
	12 +		20 ×		
				5 −	

STEP 2   And now for even more of a challenge …

Puzzle 3 (left):

3 ÷	336 ×		1 −	2 −		1 −
				17 +		
6 −		24 ×	1 −		1 −	
17 +			35 ×			
		21 ×		4 ×		3 −
	11 +		14 +			
5	6 −			36 ×		

Puzzle 4 (right):

210 ×	72 ×			11 +	4 −
	105 ×		5		2 ÷
				1 −		3 −
13 +	2 ÷		18 +			15 +
		15 +		3 −	
6 ×			14 +		
		15 +			4

👤 Activity 3   Angle relationships

You can use the same type of thinking that you applied to the Calcudoku puzzles throughout mathematics.
In this activity, you will use logical reasoning to determine the sizes of unknown angles.

STEP 1   Look at the diagram below. Then answer the questions.

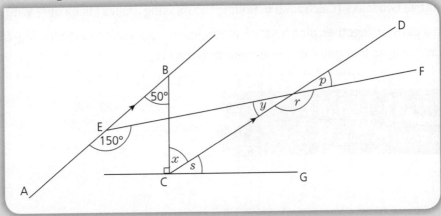

a) [State] what the arrows on the lines indicate. [Explain] how this information is useful.

b) The small italic letters represent the sizes of the angles. From the given information, which lettered
angle is easiest to determine? [Explain] your answer.

c) [Describe] the process you used to determine its size.

d) [Find] the sizes of all of the angles represented by letters. [Describe] the process you used for
each angle.

It is sometimes possible to work out the size of an unknown angle even when no other angles are given. Work out the value of x in the diagram below.

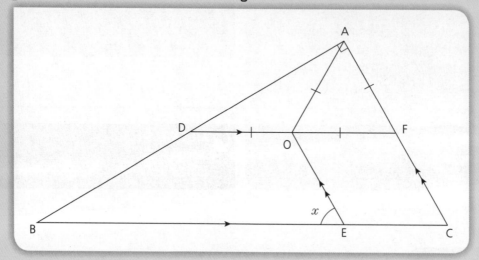

Describe the process you followed to find the value of x. **Justify** each step.

REFLECTION

a) How is the process you used to solve Calcudoku puzzles the same as the one you used to find the sizes of the missing angles? How is it different? **Explain** your answer.

b) Which of these two tasks (Calcudoku or finding the missing angles) is easier? Why?

c) Work with a partner. Together, plan a set of strategies to help solve these missing-angle puzzles. Write them down, so that you can share them with the class.

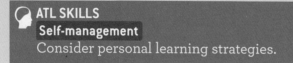

ATL SKILLS
Self-management
Consider personal learning strategies.

Summary

Logic is a method of systematic reasoning. It is used in all areas of mathematics. The puzzles in this chapter may not, at first, seem to have much formal mathematical content. However, you have used logic to solve them, and this same process of logical deduction can be applied to most mathematical topics. You will use the same skills you have used in solving these puzzles as you look for patterns or relationships and establish general rules across all areas of mathematics. As you work through this book, you will use logic to solve problems and generate rules. You might be surprised that, as you improve your skills in logical thought, mathematics seems to get easier! After all, practice makes perfect.

Introducing key concept 3: relationships

INQUIRY QUESTIONS	■ **Are relationships beneficial?** ■ **If a relationship between two things can be established, are they necessarily related?**
SKILLS	**ATL** ✓ Organize and depict information logically. ✓ Draw reasonable conclusions and generalizations. **Geometry** ✓ Solve proportions that arise from similar triangles. ✓ Establish relationships between lengths of sides of triangles. **Statistics** ✓ Use a GDC to graph data and perform a regression analysis. ✓ Interpret the relationship between two variables.

GLOSSARY

GDC graphic display calculator.

Similar the property of having the same shape, but not necessarily the same size.

COMMAND TERMS

Identify provide an answer from a number of possibilities. Recognize and state briefly a distinguishing fact or feature.

Justify give valid reasons or evidence to support an answer or conclusion.

Solve obtain the answer(s) using algebraic and/or numerical and/or graphical methods.

Suggest propose a solution, hypothesis or other possible answer.

Introducing relationships

The key concept of relationships has many applications in mathematics. It helps you to identify connections between quantities, properties or ideas. It ranges from the equivalence between decimals, ratios and percentages to how two variables relate to one another—and beyond. You can use these connections to propose models, rules and mathematical statements. Establishing such relationships is fundamental to looking at the world around you and describing the patterns that you observe. Known relationships also enable you to determine and develop new relationships. In every branch of mathematics, you will find relationships and then use them to make predictions or determine the values of variables. In this chapter you will see examples of what can happen when you can identify and then apply such relationships.

The laws of nature should be expressed in beautiful equations.

Paul Dirac

QUICK THINK

What relationships have you already learned about in mathematics? Find examples that can be expressed as equations or formulas from each branch of mathematics:

- Number
- Algebra
- Geometry and trigonometry
- Statistics and probability.

Give two examples of relationships that you have learned in mathematics that are not given in the form of an equation or formula.

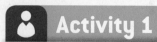
You may be surprised to learn that the study and treatment of tumours involves the use of relationships established in geometry, about **similar** triangles. Radiation therapy has been used successfully to treat tumours, but the risk of overexposure must always be kept to a minimum, if not altogether eliminated. This requires the application of mathematics to position equipment in just the right spot.

Suppose a patient needs to have their spinal cord treated with radiation that will come from two sources, as shown in the diagram below.

WEB LINKS

To see a demonstration of the relationship between similar triangles, go to http://www.mathopenref.com and search for "similar triangles".

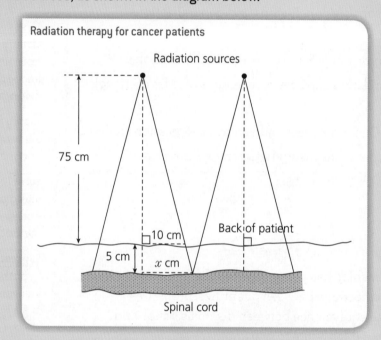

Radiation therapy for cancer patients

Focus on **one** of the two large triangles and answer the questions.

a) **Identify** the similar triangles within the larger triangle on the left-hand side. **Justify** your statement that they are similar.

b) Label the vertices of each of the similar triangles and write a mathematical statement (equation) that describes which triangles are similar.

c) Based on your answer to b), state what you know about the lengths of the sides of the two triangles that you have identified as being similar. Do some research if necessary. Express this relationship mathematically. Use an equation.

d) Use the dimensions in the diagram and your relationship from c), to **solve** for the value of x.

e) Based on your answer to d), determine how far the radiation sources should be placed from each other in order to irradiate the spinal cord without giving any section a double dose of radiation.

a) You have considered this scenario as a two-dimensional problem but, in real life, the radiation will spread out from the source in a conical formation. What implications, if any, might this have on the problem?

b) State any assumptions you made when solving this problem. Are there other elements or factors that need to be taken into account? If so, what are they and how do they affect your solution?

GLOBAL CONTEXTS
Scientific and technical innovation

ATL SKILLS
Communication
Organize and depict information logically.

 **Activity 2**

Mathematical relationships in nature— where do babies come from?

There was an ancient European legend that stated that a baby is delivered to a family in a basket carried by a white stork. In many families, that is how parents would answer the question, "Where do babies come from?" However, despite it being a legend, is there any truth in it?

In the table below, you will find the numbers of pairs of nesting storks and the numbers of out-of-hospital births that occurred in a European city for the first ten years of the 21st century.

Year	Number of pairs of storks	Number of out-of-hospital births
2000	850	889
2001	749	761
2002	745	752
2003	881	1260
2004	902	1298
2005	1089	1395
2006	1004	1324
2007	1274	1508
2008	1171	1401
2009	1202	1420

a) Graph the relationship between the numbers of pairs of nesting storks and the numbers of out-of-hospital births.

b) Describe the type of relationship suggested by the data. Express this relationship (model) mathematically.

c) Use statistics to justify your choice of model.

d) Repeat steps a), b) and c) with the following data.

Country	Storks (pairs)	Birth rate (1000/year)
Austria	300	87
Denmark	9	59
Portugal	1500	120
Romania	5000	367
Spain	8000	439
Switzerland	150	82
Turkey	25000	1576

e) How can the relationship expressed in the two tables be used to support the notion that "storks bring babies"? Is that an appropriate deduction? Explain.

f) Suggest what other explanation there could be for the simultaneous increase/decrease in the data sets.

REFLECTION

Think about the assumptions you have made and the conclusions you have drawn. Explain them to a partner.

 **GLOBAL CONTEXTS**
Personal and cultural expression

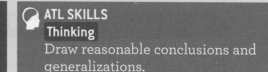 **ATL SKILLS**
Thinking
Draw reasonable conclusions and generalizations.

Summary

Relationships is a key concept that can be associated to every topic studied in mathematics. Learning to establish relationships between quantities, properties and concepts enables you to recognize patterns and construct generalizations. To communicate a relationship mathematically, you must be able to use correct mathematical language and notation, skills that also help in communicating any mathematical thinking. The skills you acquire in studying, establishing and applying mathematical relationships are central to much of what you do in mathematics.

Change

A variation in size, amount or behaviour

INQUIRY QUESTIONS

TOPIC 1 Finite patterns of change
- How can we represent change?

TOPIC 2 Infinite patterns of change
- What does being finite or infinite tell us about the size of something?

TOPIC 3 Financial mathematics
- Is it possible to predict change?

SKILLS

ATL
- ✓ Understand and use mathematical notation.
- ✓ Help others to succeed.
- ✓ Evaluate and manage risk.
- ✓ Identify trends and forecast possibilities.

Number
- ✓ Use standard form or scientific notation to represent numbers.
- ✓ Calculate percentage increase and decrease.
- ✓ Simplify results and round answers to an appropriate level of accuracy.

Algebra
- ✓ Identify arithmetic and geometric sequences and series.
- ✓ Use diagrams to represent information and solve problems.
- ✓ Calculate investment returns, using simple and compound interest.

Probability
- ✓ Calculate the probability of more than one outcome for independent events.

OTHER RELATED CONCEPTS

Pattern   Generalization   Quantity   Representation

GLOSSARY

Arithmetic sequence sequence of numbers in which each term after the first term, a, is the result of adding the same number, called the common difference, to the preceding term.

Geometric sequence sequence of numbers in which each term after the first term, a, $a \neq 0$, is found by multiplying the previous one by a number, r, called the common ratio, where $r \neq 0$, $r \neq -1$.

COMMAND TERMS

Discuss offer a considered and balanced review that includes a range of arguments, factors or hypotheses. Opinions or conclusions should be presented clearly and supported by appropriate evidence.

Explain give a detailed account including reasons or causes.

Use apply knowledge or rules to put theory into practice.

Introducing change

Change is all around you. From the population of the world to global temperatures to the life cycle of a butterfly, change occurs everywhere and in almost everything. When you started school, first you focused on things that didn't change, such as counting numbers and shapes of

Nothing is permanent except change.

Heraclitus, Greek philosopher

familiar objects. Learning about things that stay the same allowed you to measure and classify them, and count and perform operations with them. However, as you grow older, you begin to understand the constant presence of change and you therefore need to find ways to describe and understand the changes that you observe. Mathematics provides the tools to accomplish just that. This, in turn, can improve the decisions you make or even encourage you to find more creative solutions. Without change, there can be no progress.

Change represents a variation in size, amount or behaviour. Understanding variables is the first step to understanding change. Over the last few years, you have explored how changes in one variable can result in changes in another. You have also started to learn about percentage change and of rates of change, something that continues long into the study of calculus. In this chapter you will explore change in different mathematical contexts. By the end of the chapter you may find that your perspectives on a variety of situations may have changed.

🏛 MATHS THROUGH HISTORY

Over the centuries, humans have been fascinated by large numbers. Take, for example, this legend.

When the grand vizier (the highest minister in the king's court) in Persia invented chess, the king was delighted with the new game and invited the vizier to choose his own reward. The vizier replied that, being a modest man, he desired grains of wheat for his people. He took the chessboard and placed one grain of wheat on the first square, two grains on the second, four on the third, and so on, with twice as many grains on each square as on the one before. The innumerate king agreed, not realizing that the total number of grains on all 64 squares would be $2^{64} - 1$, equivalent to the world's present wheat production for 150 years! This same exponential pattern of change is what makes the game of chess itself so difficult. While there are only about 35 legal choices for each chess move, the choices multiply exponentially to yield about 10^{50} possible board positions—too many for even a computer to search exhaustively.

The grand vizier in Persia invented chess

Finite patterns of change

To start, you are going to look at two different patterns of change—arithmetic and geometric. Later in the chapter, you will look at some examples where this knowledge can be useful to help you make informed decisions.

Read this interesting story about a famous mathematician—Carl Friedrich Gauss.

🔗 CHAPTER LINKS

See Chapter 11, Pattern, for more on Gauss and his triangular numbers.

🏛 MATHS THROUGH HISTORY

Carl Friedrich Gauss (1777–1855) is considered one of the best mathematicians of all time. He worked in a wide variety of fields in both mathematics and physics, including number theory, analysis, differential geometry, geodesy, magnetism, astronomy and optics. Due to the importance of his contributions to the development of mathematics, which he called "the queen of sciences", Gauss is often referred to as the "prince of mathematics".

Supposedly, when he was a little boy, his teacher asked the class to add up the first 100 counting numbers:

$$1 + 2 + 3 + \cdots + 100$$

The teacher wanted to get some work done but to the teacher's annoyance, Gauss came up to him straight away with the answer: 5050.

QUICK THINK

How did he do that? Do these numbers have a special property?

Could he have done the same if he was to add $2 + 4 + 6 + \cdots + 200$? Or $6 + 9 + \cdots + 300$?

 Activity 1 Playing with numbers and exploring arithmetic series

STEP 1 Copy and complete this table.

Series $a_1 + a_2 + \cdots + a_n$	Number of terms (n)	Middle term	$a_1 + a_n$	Sum of the series (S)
$1 + 2 + 3$				
$1 + 2 + 3 + 4 + 5$				
$1 + 2 + 3 + 4 + 5 + 6 + 7$				
$1 + 2 + 3 + \cdots + 13 + 14 + 15$				
$1 + 2 + 3 + \cdots + 23 + 24 + 25$				
$1 + 2 + 3 + \cdots + 63 + 64 + 65$				

STEP 2 Based on the results shown in your table, write down:
- the relation between the middle term and $a_1 + a_n$
- the relation between the sum of the series S, the number of terms n and $a_1 + a_n$.

STEP 3 Write down a formula to calculate the sum of $(a+1)+(a+2)+\cdots+(a+n)$ in terms of a and n.

STEP 4 Modify your previous result to obtain a formula for the sum of consecutive terms of any arithmetic sequence. Give at least three examples. **Explain** your reasoning.

STEP 5 Note that asked to find $1 + 2 + 3 + \cdots + 100$, Gauss may have used a clever trick:

$$\begin{array}{cccccc}
1 + & 2 + & 3 + & 4 + \cdots + & 100 \\
100 + & 99 + & 98 + & 97 + \cdots + & 1 \\
\hline
101 + & 101 + & 101 + & 101 + \cdots + & 101
\end{array}$$

Then he simply noted that $101 \times 100 = 10\,100$ and the required sum could be obtained by dividing $10\,100$ by 2.

Apply your result and find, without using a calculator, a simplified expression for each of these sums.

a) $\dfrac{1}{2} + \dfrac{3}{2} + \dfrac{5}{2} + \cdots + \dfrac{99}{2}$

b) $-50 - 45 - 40 - \cdots - 5$

c) $\sqrt{2} + 2\sqrt{2} + 3\sqrt{2} + \cdots + 1000\sqrt{2}$

d) $\sqrt{2} + \sqrt{18} + \sqrt{50} + \cdots + \sqrt{882}$

STEP 6 Another version of the story about the Gauss addition trick claims that he reasoned differently: he simply paired up the hundred terms into 50 pairs with equal sum 101.

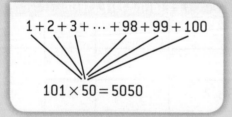

TIP

When you have a list of numbers—a numerical sequence—each number is a "term". The terms are referred to as first, second, third, etc. depending on their position in the sequence. A sequence is then defined as a set $\{a_1, a_2, \ldots\}$ or simply $\{a_n\}$ where n is the position of the term a_n. An expression obtained by adding consecutive terms of a sequence is called a series. A series may have a finite or an infinite number of terms.

How can you adapt this method to calculate the sum of an arithmetic series with an odd number of terms? For example:

$1 + 2 + 3 + \cdots + 98 + 99$

Discuss the advantages and disadvantages of each method.

⊂⊃ WEB LINKS   The video *One to One Million—Numberphile* can be found on www.youtube.com. It explores this method of adding large strings of numbers that follow an arithmetic pattern of change.

ATL SKILLS
Communication
Understand and use mathematical notation.

⊞ MATHS THROUGH HISTORY

The game of the Towers of Hanoi is based on a legend that takes several forms, but is always based on the same idea. This version is associated with a temple near Hanoi in Vietnam. The monks were instructed to solve a giant puzzle consisting of 64 golden rings, of different sizes, on a framework of three diamond needles. To solve the puzzle, they had to move the entire tower of 64 rings from the first needle to the last one. They could move only one ring at a time, and they could never place a larger ring on top of a smaller one. The legend says that when the monks finished, the world would end. How long would it take before the world ended? How many days? Years?

A simpler version, based on a framework of three spikes and three rings, can be solved in seven moves.

This legend shows how important it is to analyse patterns of change carefully, when it comes to making important decisions such as investments, borrowing money or betting.

⊂⊃ WEB LINKS
Interactive models of the Hanoi Tower can be found in the "Games" section at www.mathsisfun.com.

Visitors working on Towers of Hanoi at Universum in Ciudad Universitaria in Mexico City

Exploring the limits of a process

For many centuries, humans have tried to make their buildings as high as possible. The Pyramids in ancient Egypt and the Eiffel Tower in 1889 are well-known examples. In the 20th century, the challenge became a race: Who can build faster and highest? In the 1930s, within the span of just two years, three buildings were constructed in New York City—the Bank of Manhattan, the Chrysler Building and the Empire State Building. Each, in turn, was described as the world's tallest building.

Figure 5.1 Some of the world's tallest buildings are in New York City.

In the 21st century, the race continues with the 452 m Petronas Towers in Kuala Lumpur overtaken by the 508 m Taipei 101, and recently the Burj Khalifa in Dubai, at 828 m. This raises two questions: How high can we build? What are the limitations?

You will perform a few simple experiments and look at the relationships between the amount of material needed and height of our construction. You will deal with some big numbers and you will need to use standard form (or scientific notation) to write them down.

⊂⊃ INTERNATIONAL MATHEMATICS   Standard form (also called scientific form, scientific notation or exponential form) is a way of expressing numbers: the number is written as the product of a power of 10 and a number that is greater than or equal to 1 and less than 10.

This notation is especially useful to express very large or small numbers that astronomers, biologists, engineers, physicists and many other scientists often work with. For example, the distance of the Earth from the Sun is approximately 149 000 000 000 metres, that is, 1.49×10^{11} metres.

The use of standard form also removes possible confusion when the terminology used in different parts of the world is not the same. For example, until 1975, 1 billion had different meanings in America and Britain: 1 billion $= 1 \times 10^9$ in America but 1 billion was 1×10^{12} in Britain.

Task 1: Stacking paper

Work in a group. To complete this activity you need a pack of 500 sheets of copying paper.

A stack of paper

STEP 1   Measure the thickness of the 500 sheets and use it to estimate the thickness of a single sheet of paper. Write it down. **Use** scientific notation.

STEP 2   Estimate the thickness of a pile of a million of these sheets. Comment on the accuracy of your estimation.

STEP 3   Choose an actual skyscraper or tower and estimate the number of sheets of paper that would need to be stacked on top of each other to reach the same height as the structure you chose.

TIP

When using a calculator such as GDC or TI-Nspire, you may expect to use the number in subsequent calculations. In this case, you can store it. This is especially helpful if the number is irrational (or has a quite long decimal representation) and you want to keep the entire result for future use.

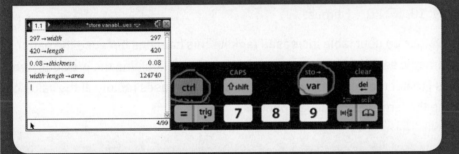

The **TI-Nspire Store Variable** operator is a secondary key, accessed by pressing **[CTRL][VAR]**. When using this method to store a variable, follow these three steps.

1. Type the **value** (or the list, matrix or expression) you want to store on the entry line.
2. Press [CTRL][VAR] to open the Store command (as indicated by a small right arrow).
3. Type the variable name and then press [ENTER] to store the variable.

The name of the variable can be a single letter or a word or letters and numbers such as *a1, length1, …*

To perform calculations with stored values just type the name of the variable in the place where you would type its value.

Task 2: Folding paper

STEP 1 Take a sheet of paper and investigate the thickness created after folding the sheet in half n times. You will need to use your result from the first question in task 1.

a) Complete the following table.

Number of folds (n)	Number of layers	Thickness
0	1	
1	2	
2		
3		
4		
5		
...		
20		
30		
40		

b) List objects or even buildings and towers with heights similar to the thickness created after folding the sheet 5, 10, 15, 20, ... times.

You may have noticed that the values on your table increased rapidly. This happens because the pattern of change is geometric. In a geometric sequence, each term is obtained by multiplying the previous term by a constant. If this constant is greater than one, the value of the terms increases rapidly. If the value of this constant is smaller than one the terms decrease rapidly.

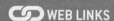 **WEB LINKS**

If you want to explore limitations of building tall towers, read the article "How tall can a Lego tower get?" available at www.bbc.co.uk/news. You can find more interesting stories about Big Numbers by searching for "Telling the Story of Big Numbers" on http://www.cre8tivegroup.com.

STEP 2 Research the maximum number of times you can fold one piece of paper. How many sheets of paper would you need to fold and then stack on top of each other to reach the same height as the Burj Khalifa (828 m)? **Discuss** the feasibility of such a construction.

 **GLOBAL CONTEXTS**
Scientific and technical innovation

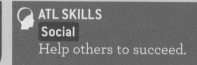 **ATL SKILLS**
Social
Help others to succeed.

Reflection

Arithmetic and geometric patterns of change are found in a wide range of applications. Investments, for example, can demonstrate either pattern. Consider the following situation:

Suppose that you have €1000 to invest and you are given two options.

Option 1: Bank A offers you 2% p.a. simple interest (so you get €20 interest every year)—so the total amount of your investment each year will be €1000, €1020, €1040, ...

Option 2: Bank B: offers you 2% p.a. compound interest, so the total amount of your investment will be €1000, €1020, €1040.40, ...

Which option represents an arithmetic pattern of change and which option represents a geometric pattern of change? Compare the amounts of your investment after 10, 20 and 50 years in each case.

TOPIC 2

Infinite patterns of change

Probability is a branch of mathematics that quantifies the likelihood of different outcomes of an event happening. The probability of an outcome happening is a number between 0 and 1 inclusive, where a probability of 0 is an impossibility, and the probability of 1 is a certainty. The larger the number, the greater is the chance that the outcome will happen.

In playing different games, you may have often wondered how high your likelihood was of winning. Depending on what happens throughout the game, your chances of winning are always changing. In the following task, you are going to investigate how changes in the initial outcomes of rolling a pair of dice change the odds of winning the dice game.

◌◌ INTERDISCIPLINARY LINKS
Probability is used in many different areas of human endeavour, for example, weather forecasting, stock-market predictions, statistics and gambling. Probability axioms are developed and defined in pure mathematics, symbolic logic and philosophy.

🏛 MATHS THROUGH HISTORY
A popular dice game in America is the game of Rolling Dice. Its origin goes back to the 12th century during the Crusades, and was eventually called Hazard in 17th and 18th century Europe. When it arrived in America, its name changed again. The game is played with two dice, and the score depends on the sum of the two dice. A player throws the two dice, and wins or loses according to the following rules:

- The player wins immediately if the total for the first throw is 7 or 11.
- The player loses immediately if the total is 2, 3 or 12.
- For any other sum (4, 5, 6, 8, 9, 10), the player scores a point.
- If the player scores a point in the first throw, they continue to throw the dice until they win by either throwing their point again (the same sum), or lose by throwing a sum of 7.

Activity 3 — Calculating the probability of winning an endless game

In this activity you will investigate your chances of winning at the game of Rolling Dice. Before completing this activity, make sure you read the rules of the game, on the previous page.

STEP 1 Draw a grid showing the possible sums when throwing two dice, one red and one blue, like this one.

	1	2	3	4	5	6
1	2	3	4	5	6	7
2	3					
3						
4						
5						
6						

STEP 2 Find the probabilities for the different sums of the dice, that is $P(\text{sum}=2)$, $P(\text{sum}=3)$, and so on.

STEP 3 Find the probability that you win on the first throw.

STEP 4 Find the probability that you lose on the first throw.

STEP 5 Find the probability that you score a point on the first throw.

STEP 6 For all throws after the first one, you will need to consider the probability of scoring the point again, that is, throwing the same sum as you did the first time in order to win the game.

Suppose you score a point for 5, that is, the sum on the first throw was a 5.

Consider the probability of winning the game on the second throw. This means throwing a sum of 5 on the second throw.

$$P(5 \text{ on 1st throw and 5 on 2nd throw}) = \frac{4}{36} \times \frac{4}{36} = \frac{1}{9} \times \frac{1}{9} = \left(\frac{1}{9}\right)^2 = \frac{1}{81}$$

STEP 7 Consider the probability of winning the game on the third throw. This means not throwing a sum of 5 (or 7) on the second throw, but scoring a sum of 5 on the third throw.

Since there are four ways of scoring a sum of 5 and six ways of scoring a sum of 7, the probability of scoring a sum of 5 or 7 is $\frac{10}{36}$. Therefore, the probability of scoring a sum of neither 5 nor 7 is $\frac{26}{36}$. Hence,

$P(5 \text{ on the 1st throw, and not a 5 or 7 on the 2nd throw, and a 5 on the 3rd throw})$
$$= \frac{4}{36} \times \frac{26}{36} \times \frac{4}{36} = \left(\frac{1}{9}\right)^2 \times \frac{13}{18}$$

STEP 8   Consider the probability of winning the game on the fourth throw. This means you would throw a 5 followed by not 5 or 7 on the second throw, not 5 or 7 on the third throw, then a 5 on the fourth throw.

$$P(5, \text{ then not 5 or 7, then not 5 or 7, then 5}) = \frac{4}{36} \times \frac{26}{36} \times \frac{26}{36} \times \frac{4}{36} = \left(\frac{1}{9}\right)^2 \times \left(\frac{13}{18}\right)^2$$

STEP 9   Find the probability of winning the game on the fifth throw. You will probably now observe that:

$P(5, \text{ then not 5 and not 7, then not 5 and not 7, then not 5 and not 7, then a 5})$

$$= \frac{4}{36} \times \frac{26}{36} \times \frac{26}{36} \times \frac{26}{36} \times \frac{4}{36} = \left(\frac{1}{9}\right)^2 \times \left(\frac{13}{18}\right)^3.$$

Make a table to summarize your results, and then complete the last two entries.

P(scoring a point on the 1st throw)	P(winning on the 2nd throw)	P(winning on the 3rd throw)	P(winning on the 4th throw)	P(winning on the 5th throw)	P(winning on the 6th throw)	P(winning on the 7th throw)
$\left(\frac{1}{9}\right)$	$\left(\frac{1}{9}\right)^2 = \frac{1}{81}$	$\left(\frac{1}{9}\right)^2 \times \frac{13}{18}$	$\left(\frac{1}{9}\right)^2 \times \left(\frac{13}{18}\right)^2$	$\left(\frac{1}{9}\right)^2 \times \left(\frac{13}{18}\right)^3$		

STEP 10   Look at your pattern in the table above. Find a formula for the probability of winning the game on the nth throw.

STEP 11   Now find the probability of winning the game **after** n throws.

TIP

You want the probability of winning after the 2nd throw, or after the 3rd throw, or after the nth throw. Deduce the probability of winning the game **after** an endless number of throws.

STEP 12   Now, repeat steps 6–11 for the different points that can be scored on the first throw. Making a table for each point would help you see a pattern to your results.

STEP 13   Which of the numbers that constitute throwing to score a point would give the higher probability of winning the game?

STEP 14   Analyse your answers to Activity 3. How many different ways are there to win the game? Generalize your results and determine the overall probability of winning the game.

REFLECTION

     a)   Based on your results, would you play the game Rolling Dice? Explain.

     b)   Do you think a game like this is fair? Explain.

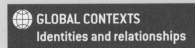
GLOBAL CONTEXTS
Identities and relationships

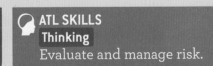
ATL SKILLS
Thinking
Evaluate and manage risk.

Financial mathematics

Financial mathematics involves creating and analysing mathematical models that represent different aspects in the financial world. You can try to make predictions about different financial trends, based on these models. This helps to make informed decisions that affect your everyday life. As in the study of probability in topic 1, the skill is in using the right mathematical tools to make good decisions.

Investment decisions

Before you invest money you should analyse all the possibilities, to see which will give the best returns (the amount of money gained by the investment). You should also consider how changing your investment strategy will maximize your investment.

For the activities in this section, assume you have US$50 000 to invest. You may choose to work with 50 000 of your national or preferred currency. You plan to invest the money for 10 years. Would you put it in the stock market, invest in bonds or put it in the bank?

Option 1: The stock market

You can buy and sell stocks of a company on the stock market. Your goal is to sell the stocks at a higher price than you bought them for. The price of each stock is determined by the supply (amount available) and demand (amount that people want) for it. When more people want to buy (demand) than sell (supply), the price of the stock goes up, and vice versa.

> 🏛 **MATHS THROUGH HISTORY**
> The first unofficial stock exchanges, in which government debt and securities were traded, took place in coffee houses in France during the 12th century. The Amsterdam Stock Exchange opened in 1602 to become the first official stock exchange. The world's largest stock exchange is the New York Stock Exchange, dating back to the late 18th century. The average daily trading value is approximately 160 billion dollars!

By looking at the past performance of the stock market you can determine the average annual returns gained (or lost).

A stock market index consists of a large group of stocks that all trade within a region (commonly by country or continent but some are global) or consist of stocks from the same sector of the market (such as technology companies). They are often used to see how the overall market is doing and as a baseline for comparison against individual stocks, portfolios and even each other.

In order to find out how the stock market has performed you will have to select a stock index (justify your choice of index) and research its current value and historic performance.

⊂⊃ WEB LINKS
You can look up historical prices of major world indices on any of the Yahoo finance sites—click on the investing section and then select world indices to find the index you are looking for. Depending on which country's Yahoo site you are on, you may have to click on "market stats" in the investing section to find the world indices.

 Activity 4   **Investing in the stock market**

To find the average annual gain or loss of the stock market, you need to pick two dates and determine the percentage by which the market has increased or decreased. Your starting date will be January 1993 and your end date will be today.

STEP 1   Find the value of the stock market on the following dates.

Starting point—1 January 1993 (historic value)

End point—today

STEP 2   **Use** this information to calculate the percentage change of the index (the market).

STEP 3   Determine the average annual gain or loss of the market (as a percentage).

STEP 4   Based on an investment of $50 000 for 10 years, what can you expect to gain or lose if you had invested in the stock market on 1 January 1993?

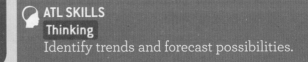

⊕ GLOBAL CONTEXTS
Identities and relationships

◉ ATL SKILLS
Thinking
Identify trends and forecast possibilities.

Option 2: Other investment types

Suppose you take your $50 000 and invest it in something other than the stock market.

Your choices are:

- investing in a 10-year government bond of your country (these types of bond generally earn simple interest)
- investing in a savings account with compound interest.

Activity 5 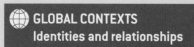 Investing in bonds

Write brief answers to these questions. You may need to do some research.

a) What is a government bond?

b) Is this generally considered a safer or a riskier investment than investing in the stock market? Why?

c) Consider the rating the government bond has been given by rating agencies. What does this rating mean, and who assigns the rating?

d) What are the advantages and disadvantages of investing in this type of bond?

e) Research the current 10-year government bond rate in your country (or the country where the majority of your stock index companies are located).

f) Calculate how much money you would make on the bond market in 10 years, if you invested in bonds instead of the stock market.

🌐 **GLOBAL CONTEXTS**
Identities and relationships

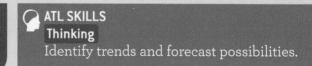 **ATL SKILLS**
Thinking
Identify trends and forecast possibilities.

 Activity 6 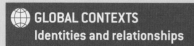 Investing in a savings account

STEP 1 What are the advantages and disadvantages of a savings account in comparison to the other investment types?

STEP 2 Research the current savings accounts that your local bank is offering.

STEP 3 Calculate the amount your savings account would hold at the end of 10 years if you had invested on 1 January 1993.

STEP 4 Compare the potential money made in all three investments in activities 4–6. Which would you choose, and why?

STEP 5 You could also decide to split the $50 000 among two or three different types of investment. What percentage of your money would you put into each type of investment? Justify your choices.

🌐 **GLOBAL CONTEXTS**
Identities and relationships

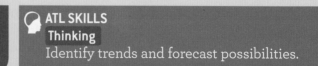 **ATL SKILLS**
Thinking
Identify trends and forecast possibilities.

You hear in the news every day about stock market volatility and bull and bear markets. In your own words, **explain** the difference between the two.

Bull and bear outside the Frankfurt Stock Exchange

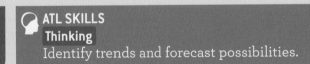

INTERDISCIPLINARY LINKS   Individuals and societies: 29 October 1929 is considered to be the worst day in stock market history. On "Black Tuesday", the stock market crashed to an all-time low and many people lost their entire life savings. It shattered the world economy and signalled the beginning of the Great Depression.

Understanding stock market trends is the key to making money on the stock market. The stock market goes through cycles that are related to the health of the economy as a whole.

Let's look at the stock market at a different point in one of these cycles.

STEP 1   New starting point—1 March 2000.
Re-do the calculations you did in Activity 1 to determine the new average annual gain or loss of the stock market.

STEP 2   Now compare your new calculation to your other two options: bonds and savings accounts. Does this change your decision?
- Does the point when you entered the market make a difference in your returns?
- Are there other factors to consider when looking at stock market returns that the index does not reflect?

 **GLOBAL CONTEXTS**
Identities and relationships

ATL SKILLS
Thinking
Identify trends and forecast possibilities.

Reflection

As a recap to this topic, try to prioritize the factors that you must take into account when investing money. Justify the order of the factors in your list.

Summary

Understanding and describing change is a central idea in mathematics. It has applications in a wide range of situations. You have used **arithmetic and geometric sequences** to describe changes of a quantity, learned about the differences between different patterns of change, and then applied your knowledge to probability and financial mathematics to help make decisions and predictions.

Equivalence

The state of being identically equal or interchangeable, applied to statements, quantities or expressions

INQUIRY QUESTIONS

TOPIC 1 Equivalence, equality and congruence
- How is it possible to be different yet equivalent?

TOPIC 2 Equivalent expressions and equivalent equations
- If something is transformed, can it still be the same as the original?

TOPIC 3 Equivalent methods and forms
- How can different methods produce the same results?

SKILLS

ATL

✓ Negotiate ideas and knowledge with peers and teachers.

✓ Recognize unstated assumptions and bias.

✓ Collect and analyse data to identify solutions and make informed decisions.

✓ Evaluate evidence and arguments.

✓ Apply skills and knowledge in unfamiliar situations.

✓ Make inferences and draw conclusions.

✓ Compare conceptual understanding across multiple subject groups and disciplines.

Number

✓ Convert between equivalent fractions.

✓ Use exchange rates to calculate values between currencies.

Algebra

✓ Rearrange equations or solve them for a variable.

✓ Determine linear equations to represent real-life scenarios.

✓ Find point(s) of intersection of linear systems of equations.

✓ Graph and factorize quadratic equations.

✓ Compare different forms of a quadratic function to determine the best to use in a given context.

Geometry and trigonometry

✓ Determine triangle congruence by proving conjectures.

✓ Use Pythagoras' theorem, the cosine rule and volume formulas to solve real-life problems.

OTHER RELATED CONCEPTS

Representation **Justification** **Simplification** **System**

GLOSSARY

Congruence equal in size and shape. The symbol for congruence is ≅.

Exchange rate the rate at which one currency can be exchanged for another.

COMMAND TERMS

Explain give a detailed account including reasons or causes.

Justify give valid reasons or evidence to support an answer or conclusion.

Solve obtain the answer(s) using algebraic and/or numerical and/or graphical methods.

Introducing equivalence

The concept of equivalence is fundamental in mathematics. You have been using it for many years, sometimes without even noticing it. The very first idea about equivalence was introduced in your first years of school: two things that are the same in some sense. For example, $2 + 3$ and $1 + 4$ are two numerical expressions with the same value. These expressions are equal. Equality is a form of equivalence.

Later on in school you learned that two circles with diameters of the same length have the same size and shape and they were called congruent. Congruence is another form of equivalence. You may have even learned about equivalent sets: the set $A = \{a, b, c\}$ and $B = \{1, 2, 3\}$ have the same number of elements. These sets are equivalent but not equal. They have something in common: the same cardinality (number of elements), although their elements are different. These examples show that the concept of equivalence is a subtle idea. Being equivalent is not necessarily the same as being equal. Equality is just a simple example of equivalence.

In this chapter you are going to explore different aspects of the concept of equivalence. In topic 1 you will meet tasks that bring you back to those definitions of equivalence as it relates to equality and congruence. In topic 2 you will encounter activities that will ask you to reflect about the meaning of equivalence in algebra: What are equivalent expressions? What are equivalent equations? You will also reflect on the advantages of replacing expressions by equivalent ones when solving problems. Topic 3 will then challenge you to use equivalent methods and forms to solve problems.

> *It is every man's obligation to put back into the world at least the equivalent of what he takes out of it.*
>
> Albert Einstein

TOPIC 1

Equivalence, equality and congruence

In this topic you will explore the concept of equivalence in terms of equality and congruence. You will start with a practical application of the concept of basic equivalent fractions that you already know, using exchange rates to calculate values between currencies.

An **exchange rate** shows amounts of money, in two different currencies, that are considered equal in value. However, do the same goods and services in two different countries cost equal amounts, even when exchange rates are used? The following activity uses the Big Mac index as a tool for comparison.

 **Activity 1**   **Big Mac index**

The Big Mac index was invented by the *Economist* magazine in 1986 as a light-hearted way to see whether a country's exchange rate is appropriate. It is based on the theory of purchasing-power parity (PPP), which says that the exchange rates should adjust to the rate that would equalize the price of an identical basket of goods and services in two countries. In this index, the basket of goods contains only the Big Mac.

∞ INTERNATIONAL MATHEMATICS
The "Big Mac" burger, made by McDonald's, is served in over 100 countries around the world and uses basically the same ingredients everywhere (except for India, which uses chicken instead of beef).

For example, to compare the United States and Russia, you look at the average price of a Big Mac in each country. In America, in July 2013, the Big Mac cost $4.56. In Russia it cost US$2.64. So the "raw" Big Mac index shows that the Russian rouble was undervalued by 42% at that time. Therefore, over time, the Russian rouble should appreciate against the US dollar in order for the Big Mac to have an equal value in each country.

Work in a group of three or four (as instructed by your teacher). Use the Big Mac index to determine if your country's currency is undervalued or overvalued, compared to those of other countries.

STEP 1   **a)** Pick five countries that you want to compare to your country. Justify why you chose those countries.

b) Find the price of a Big Mac in those countries on the list provided by your teacher.

c) Find the current exchange rates between your country and the five other countries (do a search on the internet for exchange rate conversions—there are many to choose from).

STEP 2   Use the idea of PPP to compare two currencies at a time. Calculate the percentage by which the other currencies are overvalued or undervalued, compared to your own country's currency.

REFLECTION   **Group/class discussion**

a) Based on your results, do you think your country's currency will appreciate or depreciate in the future?

b) Is the Big Mac an appropriate product to use to compare world currencies? Explain your reasoning.

c) Find a product or service that has an equivalent value in your currency and that of another country (taking into account the exchange rate). Would this be a better tool than the Big Mac index? Explain why.

 **GLOBAL CONTEXTS**
Globalization and sustainability

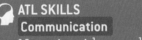 **ATL SKILLS**
Communication
Negotiate ideas and knowledge with peers and teachers.

In the next two activities, you will study the concept of congruency and analyse the use of congruent triangles in buildings and structures.

Two objects are congruent if they are exactly the same size and shape. Congruent polygons have the same number of sides as each other. All of the corresponding sides are equal and all of the corresponding angles are equal. They are still congruent, even if one of them is rotated and/or reflected, relative to the other. In fact, for all regular shapes—squares, regular pentagons or circles—it is enough to establish that they have the same size (the same side length or the same radius) in order to say they are congruent. In your study of geometry, you have established specific criteria that also help determine if two triangles are congruent. In the following activity you will be asked to use this knowledge and apply it to a special type of triangle: medial triangles.

 Activity 2   **Medial triangles and congruency**

a) Use GSP, or any dynamic geometry software to construct a triangle.

b) Construct the midpoints of each of the sides. Join up these midpoints to form a triangle inside your original triangle. This is called the medial triangle.

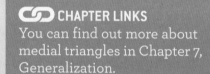
CHAPTER LINKS
You can find out more about medial triangles in Chapter 7, Generalization.

c) Determine if the four small triangles are congruent. **Justify** your conclusion, making sure you prove any conjectures you made.

d) Draw a diagram of your triangles, clearly labelling which angles and sides are equal.

GLOBAL CONTEXTS
Scientific and technical innovation

ATL SKILLS
Communication
Make inferences and draw conclusions.

In the next activity you will use the criteria for the congruency of triangles in a real-life context and then reflect on how the properties of equivalent shapes help architects and engineers make decisions about the structures they design.

 Activity 3   **Congruent triangles in structures**

STEP 1 The side lengths of triangular faces that make the apex of the Great Pyramid of Giza in Egypt are 219 m and the angle at the top (inscribed angle) is 63°.

a) Knowing that each triangular face has these dimensions, state the postulate or condition for **congruence**, such as side, angle, side, you would use to justify the congruence.

b) Ignoring the chambers dug out of the stone, find the approximate volume of stone used to build the great pyramid.

c) Choose four other three-dimensional shapes and determine their possible dimensions. Assume that the base area and the volume of stone needed to build each shape would be the same as those of the great pyramid. Why do you think the ancient Egyptians chose a pyramid as the shape for their tombs?

QUICK THINK
The great pyramids were built around 2550 BC and despite the lack of building technology, the Egyptians were able to build them with unbelievable precision. Research how the pyramids were built.

TOPIC 2

Equivalent expressions and equivalent equations

In mathematics you often study topics in which there is a systematic general method to solve problems. However, if you want to solve equations in a systematic way and look for general methods you will quickly discover that there isn't one that allows you to solve all types of equation. In fact, there is no general method that allows you to solve polynomial equations of degree above 4. To understand this problem better, look at the methods you use to solve linear equations and at the reasons why the same method cannot be extended to higher degree polynomial equations.

In earlier years you learned to solve linear equations. In fact you can solve any linear equation by always using the same principles or rules.

Addition principle: When the same value or expression is added to both sides of an equation the result is a new equation equivalent to the original one.

Multiplication principle: When both sides of an equation are multiplied by the same non-zero value or expression the result is a new equation equivalent to the original one, provided that the value is not zero.

Substitution principle: If part of an equation is replaced by an equivalent expression an equivalent equation is obtained.

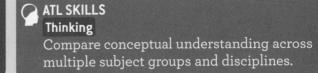

TIP

Subtraction is the addition of the opposite, and division is the multiplication of the reciprocal, so there is no need for separate principles.

An equivalence transformation consists of the application of one or more of the three principles discussed above.

By applying these rules over and over again, you obtain sequences of equivalent equations that get simpler and simpler until eventually you reach an equation for which the solution is obvious.

You can solve any linear equation by using these principles routinely.

For example, you can use equivalence transformations, like this:

- $\frac{x+1}{2}+1=\frac{5}{6}(x+2)$ is equivalent to $\frac{x+1}{2}+1=\frac{5}{6}x+\frac{10}{6}$ by the substitution principle that allows you to remove brackets (parentheses)

- $\frac{x+1}{2}+1=\frac{5}{6}x+\frac{10}{6}$ is equivalent to $6\left(\frac{x+1}{2}+1\right)=6\left(\frac{5}{6}x+\frac{10}{6}\right)$ by the multiplication principle

- $6\left(\frac{x+1}{2}+1\right)=6\left(\frac{5}{6}x+\frac{10}{6}\right)$ is equivalent to $3x+3+6=5x+10$ by the substitution principle

- $3x+3+6=5x+10$ is equivalent to $3x-5x+3-3+6-6=5x-5x+10-3-6$ by the addition principle and to $-2x=1$ by the substitution principle.

Finally, this equation is equivalent to $x=-\frac{1}{2}$ by the multiplication and substitution principles.

The last equation is so simple that its solution is shown!

 ## Activity 4   Solving equations

STEP 1   **Solve** each of these equations. **Justify** all the steps. Use the principles described above.

a) $2x-3=x+4$

b) $\frac{x-1}{3}=\frac{x}{2}$

c) $2(x-1)=4$

d) $2(3x-1)+3(x-1)=0$

e) $\frac{x}{8}-\frac{x}{4}=4$

> **CHAPTER LINKS**
> In Chapter 16, System you also justify steps when solving equations. In system, however, you use the properties of real numbers.

STEP 2   In other subjects, you may be given formulas to solve typical problems. A formula is simply an equation in which you can decide which variable acts as the dependent variable (the subject of the formula) and which one acts as the independent variable. To apply the formula and solve the problem, rearrange the formula and then replace the independent variables by the values provided, to determine the required value of your subject.

Study the two formulas below. You may already have encountered them, especially if you are studying physics. Determine what each variable represents and rearrange the equations to make the specified variable the subject of the equation, to solve for that variable.

a) Newton's law of universal gravitation: $F = \dfrac{Gm_1m_2}{d^2}$ (make m_1 the subject)

b) Kepler's third law of planetary motion: $\dfrac{4\pi^2}{T^2} = \dfrac{GM}{r^3}$ (make G the subject)

STEP 3 Look again at the formulas above. If you were to make d the subject in the formula for Newton's law, you would need to apply the method you learned to solve quadratic equations.

a) Explain how you would solve a quadratic equation $ax^2 + bx + c = 0$. Try to use equivalence transformations to justify all the steps. List any other results you may need to apply to justify the steps.

b) Reflect on the limitations of the methods used to solve non-linear equations and explain their impact on the care you need to use when rearranging formulas in other subjects.

STEP 4 Rearrange each of these equations to make the specified variable the subject of the equation, to solve for that variable.

a) $V = \dfrac{4\pi r^3}{3}$ (make r the subject)

b) $a^2 = b^2 + c^2 - 2bc \cos A$ (make angle A the subject)

c) Decide whether or not the equations obtained in a) and b) are equivalent to the original formulas. Give reasons for your answers.

REFLECTION

a) Reflect on the methods you have studied so far to solve equations. Give examples where you needed to use principles other than the ones in this topic to solve them, or had to check answers for extraneous solutions.

b) Reflect on the methods you have studied so far to rearrange formulas. Give examples of formulas where you need to be careful when rearranging them or where you need to impose restrictions on a variable.

c) After exploring the concept of equivalence formally in this topic, how did this impact on your ideas about equivalence? How would you define equivalence in mathematics? How would you define it in another context? State any similarities and/or differences between your definitions.

GLOBAL CONTEXTS
Scientific and technical innovation

ATL SKILLS
Thinking
Recognize unstated assumptions and bias.

Equivalent methods and forms

When you are solving a problem, you can often use different tools, algorithms or methods to arrive at the same result. For example, when you walk to school you may try different routes, but they will all take you to the same place. After trying them all out, you would consider all the different routes and try to decide whether they really are all equivalent, or if one path was more efficient than the others. You will ask yourself: "Does one path get me to my goal more quickly than the others?"

When you are devising algorithms, choosing methods of solution or deciding which form of representation to use, this last step is crucial for you to select the most efficient method of solution for the given problem.

Equivalent methods

Determining the most efficient method, and using multiple methods to solve the same problem to verify solutions, are valuable skills that you will develop and improve as your skill base increases. In particular, when analysing a linear system, you can solve the system graphically, algebraically or using matrices. In the tasks in this topic you will look at multiple equivalent methods and reflect on which is the most appropriate to use.

Systems of equations: 2-by-2 system of linear equations

Try this activity, based on a situation you are very likely to face, sooner or later.

 Activity 5   Buying a car

Suppose you were considering buying a car. You have to decide whether to buy a petrol, diesel or perhaps a hybrid version. How many years would it take to pay off the extra cost for the hybrid (or diesel) version with the savings you get on fuel economy?

STEP 1   Research a car that comes in alternative versions.

STEP 2   **a)** Determine the factors and assumptions you need to consider for this task.

b) What is being measured on the axes? What will the scale be? What units will you use?

c) What will the gradient and intercept on the vertical axis of each line represent? What units will you use?

STEP 3 **a)** Determine a cost equation for each version of the car. Show all calculations to determine the gradient and make note of any assumptions made.

b) Graph both equations on the same set of axes.

c) Use the graph to find the point of intersection (POI) of the two lines. What does it represent in the context of this question?

STEP 4 **a)** Use an algebraic method to solve the system of equations.

b) Which method was easier to use to find a solution? Why?

c) Which method gives you a better understanding of the problem? Why?

d) Given the context of the question, which method is the most appropriate to use to solve this problem? Justify your choice.

REFLECTION

a) Explain how the following would affect the individual equations of each car and the POI:

(i) Cost of regular petrol increases

(ii) Government introduces (or re-introduces) a tax incentive to buy a hybrid or diesel version of a car.

b) Are your equations truly indicative of the cost of the cars? What other factors should you look at?

c) Is cost the only factor to consider when deciding what type of car to buy?

d) Research the price of petrol (gas) and diesel in other countries/continents. How would the price differences impact consumer choices in these countries?

e) Compare the method used to solve linear equations with the methods used to solve systems of linear equations. To what extent can you use equivalence transformations to solve systems of linear equations? Do all the methods you know rely on equivalence transformations?

 GLOBAL CONTEXTS
Globalization and sustainability

 ATL SKILLS
Research
Collect and analyse data to identify solutions and make informed decisions.

⊂⊃ INTERDISCIPLINARY LINKS
Individuals and societies
Petrol prices differ, largely due to government intervention; some countries impose high petrol taxes while others subsidize the cost. Investigate the pros and cons of government intervention. In this situation, decide which form of government intervention (taxes or subsidies) you feel is better, clearly listing arguments to defend your position.

Systems of equations: 3-by-3 system of linear equations

If the problem you are solving has three variables, then you would need three equations to determine a unique solution (called a 3-by-3 system of equations). In the next activity, you will use what you know about 2-by-2 systems of linear equations to practise solving a 3-by-3 linear system. You will then apply an equivalent method to solve the system.

Activity 6 — A balanced diet

Most people would like to develop a healthy lifestyle. Did you know that you can use a system of equations to plan one? Select three foods that you eat almost every day and research how many grams of protein, fat and carbohydrates they provide to your daily diet, per serving.

WEB LINKS
See www.dietfacts.com for nutritional information on most foods.

Food	Protein (grams)	Fat (grams)	Carbohydrates (grams)
Recommended daily allowance	75	67	250

Use the nutritional information in your table, and your knowledge of systems of equations, to find how many servings of each food would meet the daily allowance (RDA) as recommended by the US Department of Agriculture (USDA).

STEP 1
a) Define your variables.
b) Write three equations relating the number of servings of each food required to meet the RDA for protein, fat and carbohydrate.
c) Use elimination to reduce your 3×3 system of equations to a 2×2 system.
d) Simplify this system to one linear equation and solve the problem.

STEP 2
a) Use an alternative method to solve this system of equations.
b) Which method do you prefer to use to solve this system of equations? **Justify** your choice.

STEP 3 Submit a report on what you found. Make sure that you include:
- a completed table
- a clear solution to the system, using different methods
- how many servings of each food that would allow you to meet the USDA's requirements.

STEP 4
a) Would it be possible to meet the daily requirements with only the foods you selected? **Explain**.
b) Would you want to meet the requirements with only the foods you selected?
c) Why are you recommended to eat from a variety of food groups? Do you eat enough of a variety? **Explain**.

REFLECTION
a) Is it possible for equivalent methods to produce different results? **Explain**.
b) When can you say that two methods are equivalent? Give reasons for your answer.

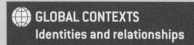
GLOBAL CONTEXTS
Identities and relationships

ATL SKILLS
Research
Collect and analyse data to identify solutions and make informed decisions.

Equivalent forms

Equivalent forms are found in many branches of mathematics. In the first two activities in this topic you will look at equivalent forms of quadratic functions and decide which is the most appropriate form to use in a given context. There are several ways to represent a quadratic function, all of which are equivalent because, for the same value of x, they all produce the same value of y.

In general, when graphing quadratic functions, there are certain points that are extremely useful:

- the vertex
- the x-intercept(s)
- the y-intercept.

Every parabola also has an axis of symmetry that goes through the vertex, as shown on the right.

Are these features easily visible in all of the forms of a quadratic function? Even though they are equivalent, are there some forms that are more useful than others? Start by looking at them individually before you use them all in a real-life application.

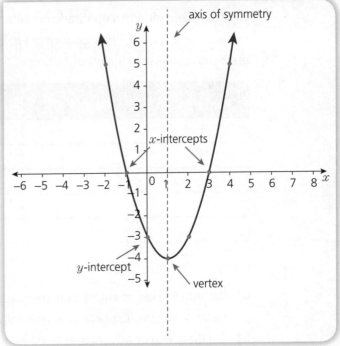

Activity 7 — Equivalent forms of quadratic functions

STEP 1 — **Standard form**

The standard form of a quadratic equation is $y = ax^2 + bx + c$, $a \neq 0$. Using a GDC, graph each function and fill in the table.

Function	Vertex	x-intercept(s)	y-intercept	Axis of symmetry
$y = x^2 + 3x - 4$				
$y = x^2 + 8x + 7$				
$y = x^2 - 7x + 12$				
$y = 2x^2 - 5x + 2$				
$y = 3x^2 + 5x - 2$				
$y = 4x^2 - 6x + 2$				

a) Identify the key features that are easiest to determine when a quadratic function is in standard form. **Explain** your reasoning.

b) How does the equation of the axis of symmetry relate to the coefficients a, b and c?

STEP 2 **Factorized form**

The factorized form of a quadratic equation is $y=a(x-b)(x-c), a\neq0$. Factorize the expressions in the table in Step 1. Write your answers in the first column in the table below. Copy your previous results into the other columns. Then answer the questions that follow.

Function	Vertex	x-intercept(s)	y-intercept	Axis of symmetry

a) Identify the key features that are easiest to determine when a quadratic function is in factorized form. **Explain** your reasoning.

b) Are there any disadvantages to working with the factorized form of a quadratic function? If there are, write them down.

STEP 3 **Vertex form**

The vertex form of a quadratic equation is $y=a(x-h)^2+k, a\neq0$. Complete the square for the functions in the table in Step 1. Write your answers in the first column below. Copy your previous results into the other columns. Then answer the questions that follow.

Function	Vertex	x-intercept(s)	y-intercept	Axis of symmetry

a) What key features become easy to determine, when a quadratic function is in vertex form? **Explain**.

b) What are the limitations to vertex form? **Explain**.

STEP 4 **Comparing the three equivalent forms**

a) Describe how you would graph a quadratic function given in standard form, without using technology. Which form do you prefer?

b) Suppose the x-intercepts are not rational numbers. Identify which form would be easiest to use to find them. Describe how you would accomplish this.

c) Suppose a fireworks manufacturer packaged the fireworks with the equation for its trajectory. Which form would you prefer it to be in? **Explain**.

d) Suppose a model of a suspension bridge has cables that can be represented by a quadratic function. Which form would be most useful if you wanted to calculate the lengths of the supports attached to the cable?

🌐 **GLOBAL CONTEXTS**
Scientific and technical innovation

💭 **ATL SKILLS**
Thinking
Evaluate evidence and arguments.

The different forms of a quadratic equation have their own advantages and disadvantages, allowing each of them to be better suited to certain problems. In the next activity, you will test your knowledge of these different forms as you apply them to a real-life situation.

 Activity 8   **Will it make it?**

In rugby, American football and Canadian football, players can score 3 points by kicking a "drop goal" or "field goal". To do this, they must kick the ball through a set of uprights/posts and over a bar, supported by the posts, that is 10 feet above the ground. While the positions of the uprights may be different in different sports, the length of the drop goal or field goal is measured from the place where it is kicked to a point directly below the uprights.

On 9 September 2012, David Akers attempted to match the current record of 63 yards for the longest field goal in American football. Assume the path of the ball can be represented by a quadratic function and that, to be successful, the ball must be at least as high as the goal post after travelling the 63 yards. Assume it's actually travelling between the posts.

Analysis of the kick found that its maximum height was around 84.5 feet.

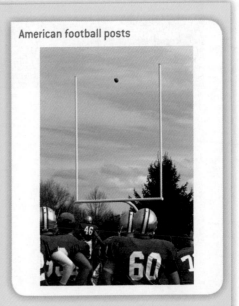

American football posts

The following data points would have also been recorded:

Horizontal distance travelled (yards)	Height above ground (feet)
0	0
20	72
65	0

a) Use the given information to find the quadratic function representing the trajectory of Akers' kick in:

   i) factorized form   ii) vertex form   iii) standard form.

b) Determine whether or not Akers ties the record. Use the equations from a), b) and c) to justify your answer.

c) Determine the furthest distance from which he could have made the kick.

d) State any assumptions you made as you solved the problem.

e) The kicker who established the record was Tom Dempsey and he did it in 1970. Research Dempsey and find out what made his record so incredible.

⊂⊃ WEB LINKS
To see the kick, go to http://www.youtube.com and search "Tom Dempsey 63 yard field goal".

REFLECTION

a) Is using an equivalent form of a function the same thing as solving a problem with an equivalent method?  Explain.

b) Explain how equivalent expressions relate to equivalent forms.

⊂⊃ INTERNATIONAL MATHEMATICS
In American football all measurements are in feet and inches rather than metres and centimetres.

⊕ GLOBAL CONTEXTS
Identities and relationships

ATL SKILLS
Thinking
Apply skills and knowledge in unfamiliar situations.

Summary

In this chapter, you explored the concept of equivalence in a variety of ways. You looked at real-life applications but also at formal principles that allow you to justify the methods you use when solving equations or working with formulas. You learned that this concept is very subtle and is present in most of the work you do in mathematics. As you continue your studies, you are now ready to notice different types of equivalence and use them to work with more suitable forms of representation or simpler expressions, or simply adopt more efficient but equivalent methods to solve problems.

Generalization

A general statement made on the basis of specific examples

INQUIRY QUESTIONS	**TOPIC 1** Number investigations
	▪ **Do patterns occur by accident?**
	TOPIC 2 Diagrams, terminology and notation
	▪ **Can all patterns be generalized?**
	TOPIC 3 Transformations and graphs
	▪ **Can all functions be transformed in the same way?**

SKILLS

ATL

✓ Use appropriate strategies for organizing complex information.

✓ Use models and simulations to explore complex systems and issues.

✓ Test generalizations and conclusions.

✓ Practise observing carefully in order to recognize problems.

✓ Draw reasonable conclusions and generalizations.

Number

✓ Discover and make generalizations about number patterns.

Algebra

✓ Use recurrence relations to determine terms of numerical sequences.

✓ Investigate the basic transformations of rational functions.

✓ Analyse and use well-defined procedures for solving complex problems.

Geometry and trigonometry

✓ Construct diagrams and use appropriate notation to explore networks to discover patterns and rules associated with them.

✓ Construct diagrams to explore the medial triangle to discover its properties.

OTHER RELATED CONCEPTS

Pattern   Representation   Justification   Space

GLOSSARY

Network a set of objects called vertices (or nodes) in which relationships are represented by segments (or edges). Networks (also called graphs) are useful to represent information and visualize connections between elements.

COMMAND TERMS

Describe give a detailed account or picture of a situation, event, pattern or process.

Identify provide an answer from a number of possibilities. Recognize and state briefly a distinguishing fact or feature.

Justify give valid reasons or evidence to support an answer or conclusion.

Introducing generalization

When the Hungarian mathematician George Pólya attended school, he was very frustrated by continually having to memorize information. He had the same problem at university, where he changed courses several times. Finally, when he decided to study mathematics and physics, he discovered that what he really enjoyed was solving problems. His first job was to tutor Gregor, the young son of a baron, who struggled in mathematics due to his lack of problem-solving skills. Pólya spent many hours developing a systematic method of problem-solving that would work for Gregor, as well as others in the same situation. Later, in 1945, he published *How to Solve It*, in which he developed problem-solving strategies in detail. This is still regarded as the most famous problem-solving book ever written.

🏛 MATHS THROUGH HISTORY

George Pólya (1887–1985) was born in Budapest, Hungary, where he completed his initial education. In high school, he did not do well in mathematics. He later blamed this on poor teaching. He had some trouble in deciding what to study at university; first he followed his mother's wish and studied law, then languages— Latin and Hungarian—and eventually mathematics. In 1911, he moved to Vienna where he studied for a year. After this, he continued studying and working at universities in Göttingen, Paris, Zurich, Oxford, Cambridge and Stanford. His work as a mathematician inspired Escher. His contributions to education still influence many teachers.

Since generalization is a particular type of problem-solving, you will work on tasks to which you can apply some of Pólya's strategies. Since Pólya first published his problem-solving strategies, other mathematicians and educators have added to them. These combined strategies are summarized in this diagram.

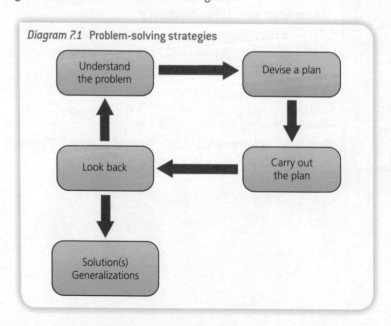

Diagram 7.1 **Problem-solving strategies**

At each stage of this process, there are specific strategies that may help find the solution. Here is a comprehensive list.

1. **Understand the problem.**

   - What data do you have? What are the variables? What are the conditions?
   - Are the conditions and data sufficient?
   - Draw a diagram or a graph. Use colours if you wish to.
   - Introduce suitable notation that is easy to remember.
   - Explore symmetry. Look for patterns.
   - Separate the parts of the problem.

2. **Devise a plan.**

   First, look at your problem.

   - Have you seen this problem before? Maybe in a different form?
   - Do you know about any related problems?
   - Do you know a theorem that might be useful?
   - Think about the techniques that you know.

   Look at the variables.

   - Find the connection between the data and the variables.
   - You may have to consider auxiliary problems or restate the problem.
   - Can you imagine an easier related problem that you can solve?
   - Can you solve part of the problem? How far can you go?
   - Can you change the data or the variables (or both)?
   - Did you use all the data? Did you use all the conditions?

3. **Carry out the plan.**

   - Check each step. Can you justify it?
   - Are there any restrictions?

4. **Look back over what you have done.**

   - Check the result. Check the arguments.
   - Can you derive the result in a different way?
   - Are there restrictions to your results?

5. **Generalizations**

   - How can you generalize your result?
   - Can the method or the general solution be used to solve other problems?

Number investigations

Now you can apply these strategies to discover a numerical pattern. The first activity guides you through the problem-solving process outlined above.

 CHAPTER LINKS
Chapter 11 has more information on this sequence of numbers.

 **Activity 1**   Climbing stairs

When Kathy climbs the stairs, she can take one or two steps at a time. Every time she climbs the stairs she tries to do it in a different way. Work out how many different ways she can reach any given stair.

a) Draw diagrams showing all the ways she can get to the first, second, third, fourth and fifth step by climbing one or two steps at a time.

b) Construct a table showing your results.

c) Introduce simple notation, such as (1, 2, 2) to represent climbing five stairs starting with one step and then two steps twice. Use this notation to list all the different ways she can get to the sixth step if she climbs one or two steps at a time.

d) Investigate further until you find a pattern that allows you to calculate the number of ways she can get to the nth step if she climbs one or two steps at a time. Assume that there is only one way of getting to step zero.

e) **Describe** the pattern. Explain why you can generalize it for any number of steps. How do you know your generalization is correct? **Justify** or validate your general rule.

f) Use your rule to determine the number of different ways she can get from the basement to the first floor, if she needs to climb 37 steps.

TIP

The order in which she climbs the steps is important: (1, 2, 2) is different from (2, 1, 2).

TIP

It is often helpful in problems such as this one to tabulate your results.

EXTENSION

a) Can your method be used to solve a different problem?

b) Repeat your investigation, assuming that Kathy can climb one, two or three steps at a time.

REFLECTION

Think again about this problem. What was the general case? What specific cases did you consider? How did looking at specific cases help you work out the general case?

ATL SKILLS
Organization
Use appropriate strategies for organizing complex information.

Now you can apply Pólya's problem-solving techniques to solve a simplified real-life problem. Remember that you may need to simplify the first steps to begin to see a pattern. You might also need to draw some diagrams and tables to help you organize your findings.

 Activity 2   **Bad apples—it takes only one to spoil the crate!**

Apples that are similar in size are placed in a square cardboard flat crate. Each apple is represented in the diagram by a circle. Each apple just touches the others that are near it.

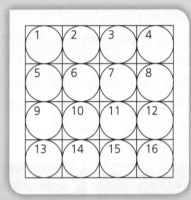

Suppose one randomly selected apple in the crate, say number 5, goes bad. How will the dimensions of the box affect how long it takes for the whole crate of apples to go bad?

Apple number 5 is the first one to go bad. One day later, all the apples that touch it (those that share a common side in the diagram) go bad. So now apples numbered 5, 1, 6 and 9 have all gone bad. After another day, the bad apples spoil all the apples they touch (those with which they share a common side). This continues until all the apples in the crate have gone bad.

STEP 1 **a)** Find out how many days it takes for the entire crate of apples to go bad.

b) Do this again, but this time with a different apple going bad first. Again, find how many days it takes for the entire crate of apples to go bad.

c) Continue until you have investigated all the possible numbers of days it would take for the entire crate of apples to go bad. Make sure you comment on the position of the first bad apple.

STEP 2 **a)** Repeat step 1, using square crates of different sizes.

b) Devise a formula to calculate the number of days it takes for an entire square crate of apples to go bad. Make sure you consider the position of the first bad apple.

c) Determine if there is a difference in times between squares with odd and even numbers of apples along the side.

STEP 3 **a)** Investigate how long it takes for all apples in rectangular crates to go bad.

b) Look at different-sized rectangles with length l and width w, and different positions for the first bad apple. Then, find a formula to calculate the time it takes for rectangular l by w crates of apples to go bad.

STEP 4 Now suppose apples were put into square boxes with crates stacked one upon the other, so that they also touch one another on the top and bottom. How long will it take for the entire box of apples to go bad if initially only one apple in the entire box is bad?

STEP 5 How long will it take for an entire rectangular box of apples, of dimensions a by b by c, to go bad if initially only one apple in the box goes bad?

REFLECTION

a) What were the specific results you found in this problem? What were the general results you found? Why is it useful to have a general result?

b) Make generalizations about how the apples have been packed and how long they will stay fresh if one starts to go bad. Research how different kinds of fruit are packed commercially. What considerations are made regarding contamination? Take into account factors such as ease of shipping, costs of shipping and costs of packing material used. Balance them against the costs of packing them to minimize contamination.

🌐 **GLOBAL CONTEXTS**
Globalization and sustainability

🧠 **ATL SKILLS**
Organization
Use appropriate strategies for organizing complex information.

Reflection

Reflect on the strategies you used to solve the activities in this topic and how you could improve your approach to problem-solving.

TOPIC 2

TIP

Diagrams, terminology and notation

In this topic, you will explore problems that highlight the importance of using simple diagrams and clear specific terminology and notation. The first is a **networks** question that comes within a branch of mathematics called *discrete mathematics*. This is a fairly new branch that has developed quickly due to its applications to computer and automated systems. Mathematicians working in this area need to be extremely careful to use clear diagrams and very specific terminology to communicate their ideas. The second problem is an analytic geometry question, in which diagrams are crucial to understanding and generalizing the properties of geometric shapes to help solve problems. After this you will have the opportunity to practise caution in making generalizations and to analyse the steps in reasoning.

Discrete mathematics provides methods to solve problems involving integers or collections of objects that can be counted. Graph theory is an important branch of discrete mathematics that uses very specific terminology. For example, the word "graph" has a different meaning in graph theory—it refers to a collection of points and arcs connecting them.

Activity 3 — Can he do it?

The postman and circuits problems

This map shows some roads on a small island. Every day the postman has to walk along every road to distribute the mail. He wonders if it is possible to do this without having to go along any road more than once.

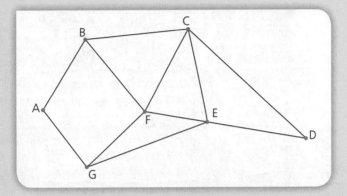

a) Can he do it? If so, where should he start and finish?

b) Look at each diagram. Try to find a possible route for the postman that goes along every road exactly once. For each diagram, write down all possible starting and finishing points. Use different colours for different routes. If you think there is no possible route, write IMPOSSIBLE.

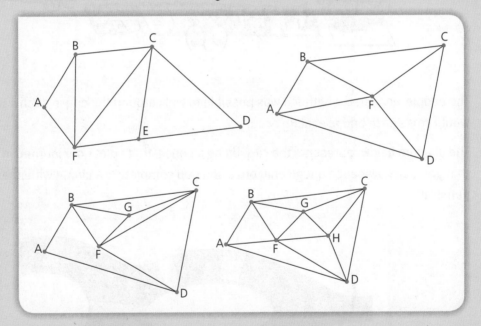

c) Look at your diagrams carefully. Try to find out what makes it possible or impossible to find a route that satisfies the condition of going along each road exactly once.

d) Look again at the networks that do have possible routes. Which ones start and finish at the same place—which ones are circuits? What do you notice about them? Now look at the routes that are possible but have different starting and ending points. What do you notice about them? Summarize your rules.

e) Sketch some diagrams of your own to test if your rule works. Generalize and write your conclusions.

Bridges of Königsberg

A river runs through the city of Königsberg, in Prussia. The city centre was an island. After passing round the island, the river broke into two parts. The people of the city built seven bridges so that they could get from one part to another. A simple map of the centre of Königsberg might look like this.

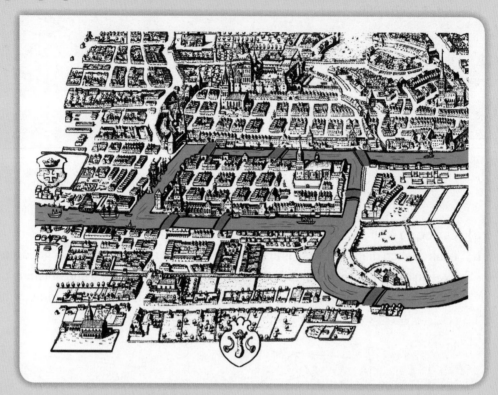

The people wondered whether it was possible to walk around the city in such a way that they would cross each bridge exactly once.

The diagram below represents the city. Using a pencil, try to plot your journey in such a way that you trace over each bridge only once and you complete the circuit with one continuous pencil line.

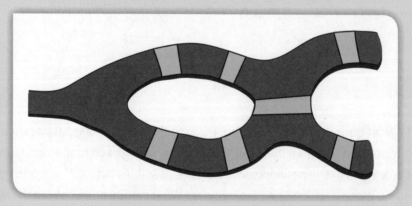

A famous mathematician, Leonard Euler (pronounced oil-er), realized that all problems of this form could be represented by replacing areas of land with points (he called them *vertices*), and the bridges to and from them with arcs. For Königsberg, he represented land with red dots and bridges with black curves:

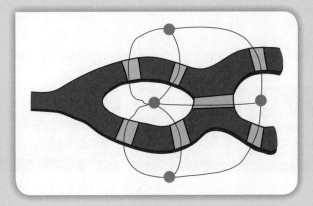

Then the seven bridges problem looks like this.

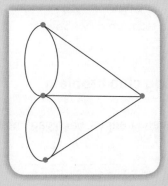

The problem is now how to draw over this picture without retracing any line and without taking your pencil off the paper. Euler generalized this method by making the following definitions.

- A graph (or network) is a figure made up of points (vertices) connected by non-intersecting lines (arcs).
- A vertex is called odd if it has an odd number of arcs (or edges) leading to it, otherwise it is called even.
- An Euler path is a continuous path that passes through every arc once and only once. When the starting and end points are the same, the Euler path is called an Euler circuit.

a) Over the years, the seven original bridges have deteriorated or been destroyed and not all have been replaced. Some of the Königsberg bridges were bombed by Allied forces during the Second World War. There are now five bridges left and it is said to be possible now to walk over all of the bridges, following an Euler path. Determine which bridges have not been rebuilt and draw a network with only the remaining bridges. You need to list the edges of your Euler path and state where you would have to start and finish your tour if you wanted to cross over each of the remaining bridges only once.

The old city of Königsberg is now called Kaliningrad (in Russia). You can check your answer by looking at a satellite image of the bridges as they stand today.

b) Did it matter which bridges were destroyed, to make it possible to walk along an Euler path and cross all of them? What if new bridges had been built? Draw some graphs of your own, showing different possible new bridges. Try to plan your journey for each graph. Write down your conclusions. Use the graph theory terminology you have learned.

Using what you learned in Activity 1, modify your strategy to complete the next activity. Use graphs to represent each situation and the terminology above to communicate your findings.

 **Activity 4**   **The pony club problem**

This is the plan for a farm with fields and hedges. Each gap shows the position of a gate.

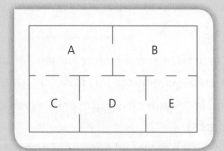

A pony club wants to make a jumping course. The course can start in any one of the five fields, A, B, C, D or E, but it must go over every gate exactly once. It can finish in any of the fields.

STEP 1 **a)** Try to find a possible route for the course. Use the diagram above to draw it.
b) Try to find another possible route. Use a different colour to draw it.
c) Can you find a course that starts in field A? And in field B? If they are possible, draw them. Otherwise, try to find a reason why it is impossible.

STEP 2 **a)** For each of the following diagrams, try to find a possible route for a jumping course that goes over each gate exactly once.
b) For each diagram, write down all possible starting and finishing fields. Use different colours for different routes. If you think there is no possible route, write IMPOSSIBLE.

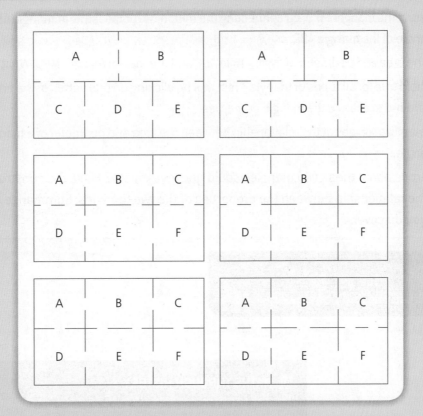

c) Look at your diagrams carefully. Try to find out what makes it possible or impossible to find a route. What makes it possible for some fields to be starting or ending points? Sketch some diagrams of your own to test if your rule works. Generalize and **justify** your conclusion.

REFLECTION

What connections can you see between Activities 3 and 4? How do Pólya's steps for problem-solving help you solve this problem quickly?

GLOBAL CONTEXTS
Orientation in space and time

ATL SKILLS
Thinking
Use models and simulations to explore complex systems and issues.

Activity 5   The medial triangle

A medial triangle is one that is constructed within another triangle, with its vertices at the midpoints of the sides of the outer triangle. To explore the properties of medial triangles, you will need to draw a diagram that you can manipulate and observe the results.

a) Use dynamic geometry software to construct a triangle ABC.

b) Construct the midpoints of each side. Label them M, P and Q.

c) Now construct the triangle MPQ, by connecting the midpoints of the sides of triangle ABC. This is the medial triangle of the triangle ABC.

d) Calculate the area and perimeter of the triangle ABC and the medial triangle MPQ. What do you notice?

e) Does this hold true for all types of triangle? Test this by clicking on one vertex of the original triangle and moving it around to change the triangle properties.

f) Make a generalization about the relationship between the area and perimeter of a triangle and its medial triangle.

What other properties about these two triangles could you investigate? Make the constructions while looking at the relationship between the properties of the two triangles. Summarize all of the generalizations you discover.

ATL SKILLS
Thinking
Test generalizations and conclusions.

🏛 **MATHS THROUGH HISTORY**
Thales (approximately 624 BC–547 BC) was a Greek engineer, although he was also known as a philosopher, scientist and mathematician. Aristotle considered him the first philosopher in the Greek tradition and he was well known for trying to explain natural phenomena without the use of mythology. His "intercept theorem" is actually evident in the preceding activity.

Beware of quick generalizations!

Mark Twain, one of the most famous early American authors, is quoted as having said, "All generalizations are false, including this one." Alexander Dumas, a famous French author, added, "All generalizations are dangerous, including even this one." What made these two people comment on generalizations in this way? What is it about words such as "all", "always" or "never" that should make us stop and think, particularly when using these words in mathematics? In Chapter 8 on justification, you will learn that in order to use such words, you must follow a rigorous process of mathematical proof.

 Activity 6   **Exercising caution**

STEP 1 **1 for all and all for 1!**
Consider this pattern.

$1 \times 1 = 1$
$11 \times 11 = 121$
$111 \times 111 = 12\,321$
$1111 \times 1111 = 1\,234\,321$

a) [Describe] the pattern that is emerging.

b) Conjecture what 11 111 × 11 111 will be. Then confirm your conjecture.

c) Find out if the pattern continues.

d) Find any limitations on the pattern.

e) Try to make a generalization about this pattern.

You will have already seen that, before making generalizations based on a pattern, you must exercise caution. You must make sure that once you have made a conjecture, you test it further on other examples.

STEP 2 **Let's circle it!**

This diagram shows circles with points on the circumferences. The number under each circle tells you how many regions the circle is divided into when the points are connected.

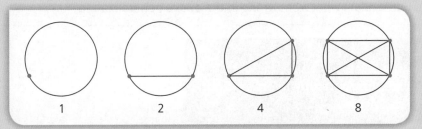

Make a conjecture for the number of regions created when there are five points on the circle, then check your conjecture. Check your conjecture for six points on the circle. Comment on your results.

STEP 3 **Proof that** $2 = \sqrt{2}$

Consider a square with sides of length of 1 unit.

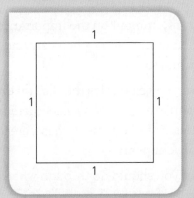

Next, divide two adjacent sides in half.

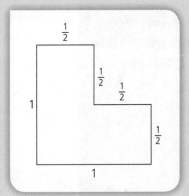

The sum of the lengths of each part making up the L-shape together with half the sides of the original square is still 2; that is, the total length is not changed.

Now, divide the two sides of length $\frac{1}{2}$ in half again, so that each one has a length of $\frac{1}{4}$.

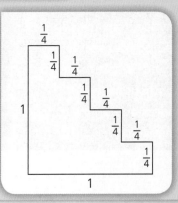

Adding all of the L-shapes together with the shortened sides of the square again gives a sum of 2.

Continue this process as shown in the diagram, and then indefinitely. Following the same reasoning, the length of the (almost) diagonal is still 2, just as the sum of the two unchanged sides of the square is 2.

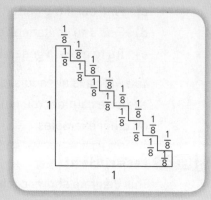

When you continue this process indefinitely, you will get a straight line, instead of the L shapes. Using Pythagoras' theorem, you know that $1^2 + 1^2 = 2$, and hence, the diagonal should be $\sqrt{2}$, but you have shown that the length of the diagonal is 2!

Can you explain what went wrong in the above generalization?

 ATL SKILLS

Thinking

Practise observing carefully in order to recognize problems.

Reflecting on diagrams, terminology and notation

a) Look again at Activities 3 and 4 and identify and list the similarities between them. Generalize your results and state the conditions that are necessary for a simple graph to be: i) an Eulerian circuit ii) an Eulerian path.

b) Look also at Activity 5. Comment on the importance of using diagrams and common terminology when carrying out activities such as 3, 4 and 5.

c) Suggest common everyday examples that use diagrams to simplify or clarify a situation. You can select them from any branch of mathematics. Based on one of these, create a task that you could give one of your peers.

d) Explain the steps that you should go through when attempting to make a generalization from an observed pattern. Include examples of where you should exercise caution. Explain the process that mathematicians must follow before a generalization is allowed in mathematics.

Transformations and graphs

In this topic, you will look at different graphs: graphs of functions. Functions are relations between two variables—usually x and y—and can be represented by sets of points on the Cartesian plane. Most functions have a domain that is either an interval or a union of intervals of real numbers. Their graphs can be drawn by using continuous lines or curves. For example, straight lines represent linear functions geometrically; quadratic functions are represented by curves called parabolas. Both linear and quadratic functions have the set of real numbers as their domain and their graphs are continuous lines or curves.

The next activity guides you through an exploration of graphs of a special family of rational functions with graphs that are represented by curves called hyperbolas. You will start with the graph of the function $y = \dfrac{1}{x}$. This is a very special hyperbola—an equilateral one that can be drawn easily as its asymptotes are the coordinate axes. Then you will be asked to use your calculator to investigate different transformations of this function and its graph. For each activity, you need to summarize your findings and write down a general statement for your results.

> **TIP**
>
> In this context, the word *graph* no longer refers to discrete sets of points.

Activity 7 — Vertical stretches

a) In the given equation $y = \dfrac{a}{x}$, choose different **positive** values of a to determine how it affects the original graph $y = \dfrac{1}{x}$.

b) How do your results change when you choose different values for a?

c) Use a GDC to graph your selected functions on the same set of axes.

d) Create a table to summarize the characteristics of each graph.

e) What happens to the graph of $y = \dfrac{1}{x}$ when the expression $\dfrac{1}{x}$ is multiplied by a positive number? Write a general statement about the effect of changes in the value of a on the graph $y = \dfrac{a}{x}$ when $a > 0$.

f) Generalize your results and write down the general statement that summarizes your findings.

ATL SKILLS
Thinking
Draw reasonable conclusions and generalizations.

Activity 8 Reflections

a) In the given equation $y = \dfrac{a}{x}$, choose different negative values of a to determine how it affects the original graph $y = \dfrac{1}{x}$.

b) [Justify] your choice of values for a. Do you think your results would be different if you chose different values for a?

c) Use a GDC to graph your selected functions on the same set of axes.

d) Create a table to summarize the characteristics of each graph.

e) What happens to the graph of $y = \dfrac{1}{x}$ when the expression $\dfrac{1}{x}$ is multiplied by a negative number? Write a general statement about the effect of changes in the value of a on the graph $y = \dfrac{a}{x}$ when $a < 0$.

f) Generalize your results and write down the general statement that summarizes your findings.

 ATL SKILLS
Thinking
Draw reasonable conclusions and generalizations.

Activity 9 Translations

a) In the given equation $y = \dfrac{1}{x} + k$, choose different values of k to determine how it affects the original graph $y = \dfrac{1}{x}$.

b) [Justify] your choice of values for k. Do you think your results would be different if you chose different values for k?

c) Use a GDC to graph your selected functions on the same set of axes.

d) Create a table to summarize the characteristics of each graph.

e) What effect does changing the value of k have on the graph $y = \dfrac{1}{x} + k$ if $k > 0$?

What effect does changing the value of k have on the graph $y = \dfrac{1}{x} + k$ if $k < 0$?

Write a general statement about the effect of changes in the value of k on the graph $y = \dfrac{1}{x} + k$.

f) Generalize your results and write down the general statement that summarizes your findings.

 ATL SKILLS
Thinking
Draw reasonable conclusions and generalizations.

Activity 10 More translations

a) In the given equation $y = \dfrac{1}{x-h}$, choose different values of h to determine how it affects the original graph $y = \dfrac{1}{x}$.

b) ⟦Justify⟧ your choice of values for h. Do you think your results would be different if you chose different values for h?

c) Use a GDC to graph your selected functions on the same set of axes.

d) For the graphs, create a table to summarize the characteristics of each graph.

e) What effect does changing the value of h have on the graph $y = \dfrac{1}{x-h}$ if $h > 0$?

What effect does changing the value of h have on the graph $y = \dfrac{1}{x-h}$ if $h < 0$?

Write a general statement about the effect of changes in the value of h on the graph $y = \dfrac{1}{x-h}$.

f) Generalize your results and write down the general statement that summarizes your findings.

 ATL SKILLS

Thinking

Draw reasonable conclusions and generalizations.

 Activity 11 Summary of generalizations

Using all of the information you discovered in Activities 7–10, create a summary of all of the different types of transformation, detailing your generalizations about each type and using sketches to illustrate your general rules.

 ATL SKILLS

Thinking

Draw reasonable conclusions and generalizations.

Create an illustration of each of the rational functions below. Use $y=\dfrac{1}{x}$ as a reference on each illustration.

$$y=-\frac{4}{x} \qquad y=\frac{2}{x}-7 \qquad y=-\frac{5}{x+3} \qquad y=\frac{1}{3(x-6)}+8$$

Explain clearly each step that is required to transform the function $y=\dfrac{1}{x}$ into each of the four separate rational functions.

Graph	Description of transformations
$y=-\dfrac{4}{x}$	
$y=\dfrac{2}{x}-7$	
$y=-\dfrac{5}{x+3}$	
$y=\dfrac{1}{3(x-6)}+8$	

Use the table format below to summarize the characteristics of these graphs.

Graph	Equation of vertical asymptote	Equation of horizontal asymptote	Domain	Range
$y=-\dfrac{4}{x}$				
$y=\dfrac{2}{x}-7$				
$y=-\dfrac{5}{x+3}$				
$y=\dfrac{1}{3(x-6)}+8$				

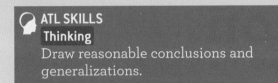

Reflecting on transformations and graphs

Write the answers to these questions in sentence form. Where it is
appropriate, use examples to illustrate your point.

a) How did the use of technology help you during the course of
this investigation?

b) How do you know your generalizations about the transformations
are correct? **Justify**/validate your general rules.

c) Can these generalizations be applied to other functions?
Explain your reasoning, showing examples of other types of
functions as validation.

Summary

In this chapter you have explored a systematic strategy to solve
problems. You have had the opportunity to apply it to different areas
of mathematics, ranging from number through to discrete
mathematics, geometry and algebra. For most tasks you confirmed
generalization of results by further testing. The activity "Exercising
caution" provided you with examples of situations where you had
to look back over your work and rethink your conclusions. The
generalization of particular results requires care—never conclude too
quickly that something is always true!

Justification

Valid reasons or evidence used to support a statement

INQUIRY QUESTIONS	**TOPIC 1** Formal justifications in mathematics

INQUIRY QUESTIONS

TOPIC 1 Formal justifications in mathematics
- What is irrefutable proof?

TOPIC 2 Empirical justifications in mathematics
- If something is justifiable, does that make it right?

TOPIC 3 Empirical justifications using algebraic methods
- What justifies the choices we make?

SKILLS

ATL
- ✓ Practise flexible thinking—develop multiple opposing, contradictory and complementary arguments.
- ✓ Interpret data.
- ✓ Consider ethical, cultural and environmental implications.

Algebra
- ✓ Use basic algebra to conduct direct proofs.
- ✓ Factorise expressions to make a proof by counter-example.
- ✓ Use linear functions and convergence of geometric series to conduct a visual proof.
- ✓ Graph a rational function and use it to make an empirical justification.
- ✓ Represent a piecewise-defined function both graphically and algebraically.

Geometry and trigonometry
- ✓ Use basic geometry and vertically opposite angles to conduct direct proofs.
- ✓ Use the properties of equilateral triangles to conduct a visual proof.
- ✓ Use circle and line theorems to conduct a visual proof.

Statistics and probability
- ✓ Use linear regression and a line of best fit to make predictions and empirical justifications.

OTHER RELATED CONCEPTS

Pattern   Generalization   Representation
Change   System

GLOSSARY

Direct proof a sequence of logical steps using only axioms and previously proved theorems to justify each step for the purpose of obtaining the required result.

Indirect proof also known as a proof by contradiction, this is a type of proof in which a statement to be proved is assumed false and then, using logical deduction, the assumption leads to an impossibility, hence the statement that was assumed to be false is proven to be true.

COMMAND TERMS

Justify give valid reasons or evidence to support an answer or conclusion.

Prove use a sequence of logical steps to obtain the required result in a formal way.

State give a specific name, value or other brief answer without explanation or calculation.

Use apply knowledge or rules to put theory into practice.

Introducing justification

Throughout this book you are solving mathematics problems, both theoretical and in a real-world context. Once you have an answer, you ask yourself if it makes sense in the context of the question. How do you know it makes sense? This is one of many forms of mathematical justification. Many teachers would argue that your ability to justify your answers is as important as the answer itself!

> *Proofs are to mathematics what spelling is to poetry. Mathematical works do consist of proofs, just as poems do consist of characters.*
>
> Vladimir Arnold

TOPIC 1

Formal justifications in mathematics

What does it actually mean to prove a mathematical statement? A proof is a series of true statements that uses both logic and previously known facts to arrive at a new truth. Mathematical proofs must be irrefutable, able to be generalized and convincing to everybody.

Proofs have been referred to as the heart of mathematics.

🏛 **MATHS THROUGH HISTORY**

The Greek mathematician Euclid, often called the father of geometry, developed a reasoning process called deduction to use for proofs in his book *The Elements*, written around 300 BC. Most of the geometry you study comes directly from this book. Euclid stated that if you have a conjecture, or proposition, that you want to prove is true, you start with self-evident facts called axioms or postulates. You put them together with other conjectures that have been proven, called theorems. Using these axioms and theorems, you create a sequence of logical arguments that prove that your mathematical conjecture is correct. This is true for all ares of mathematics, from simple algebra or geometry, to complex proofs, using many branches of mathematics and running to many pages.

Figure 8.1 Statue of Euclid in the Oxford University Museum of Natural History

STEP 1   **Direct proof using geometry**

Using the theorem that the sum of the angles on a straight line is equal to 180°, **prove** that vertically opposite angles are always equal, that is, $A = B$ and $C = D$.

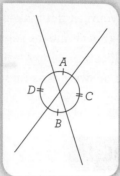

When Euclid finished his proofs, he would write QED at the end. This stands for the Latin phrase *quod erat demonstrandum,* which means "which was to be demonstrated".

In step 1, you could have drawn and measured any number of sets of vertically opposite angles, all different, and you would have found that vertically opposite angles are always equal. You could repeat this experiment in different classrooms all over the world; you would always find that the vertically opposite angles are equal. Is this enough for a mathematician to conclude that all pairs of vertically opposite angles are equal? The answer is "No!" You may arrive at a conjecture through repeated experiments, but you cannot use experimentation to justify your conjecture. In mathematics, conjectures can only become theorems through the rigorous process of deductive reasoning.

🏛 MATHS THROUGH HISTORY

For centuries, Earth was viewed as the centre of the universe. The theory was based on evidence, such as the stars appearing to move around the Earth and the fact that people couldn't really feel the Earth move. This view was popularized by Aristotle. Later Claudius Ptolemaeus improved upon this theory by adding in circular orbits that helped explain why some planets appear to move backwards in the night sky. Without telescopes or more modern tools, the justifications made sense. However, Nicolaus Copernicus eventually proposed that a heliocentric system, with the Sun near the centre of the universe, would also explain phenomena such as the changing of the seasons. Despite his mathematical justifications, even his theory underwent improvements by Johannes Kepler who analysed data on planetary motion to reveal that the orbits of the planets are elliptical, not circular.

Our solar system

STEP 2   **A direct proof, using algebra**

a)   Add several pairs of odd numbers. What do you notice?

To formalize a **direct proof**, the best approach is to follow the steps:

$$\text{experiment} \rightarrow \text{conjecture} \rightarrow \text{prove.}$$

A conjecture is a general statement that you believe to be true but have not yet proved. Make a conjecture, based on what you have noticed.

Based on experimentation, you could conjecture that the sum of two odd numbers is always an even number. However, as you know, experimentation is not a valid process for justifying a conjecture. Therefore, how can you use algebra and deductive reasoning to prove this statement?

"The sum of two odd numbers is an even number."

Begin by defining an even number algebraically. Any even number can be written in the form $2a$ where a is a positive integer.

It therefore follows that any odd number can be written in the form $2a - 1$.

$\boxed{\text{Use}}$ these definitions to express two odd numbers m and n algebraically, and $\boxed{\text{prove}}$ the conjecture.

b)   Square any even number and $\boxed{\text{state}}$ what kind of a number you get. Square any odd number, and $\boxed{\text{state}}$ what kind of a number you get. Formulate two conjectures and $\boxed{\text{prove}}$ them algebraically. $\boxed{\text{Use}}$ the definitions of odd and even numbers given in **a**.

STEP 3   **Indirect proof** (proof by contradiction)

Sometimes it is easier to prove a statement indirectly rather than using the direct method. In this method, you assume what you want to prove is not true then, using deduction, arrive at an impossible conclusion. This means that what you assumed is not true, so therefore the opposite must be true.

You have proven in step 2, part **b**, that if you square an even number, the result is even. Is the converse of this statement true? If the square of a number is even, is the original number even?

a)   Experiment with different examples, and make a conjecture.

b)   Now assume that there exists an odd number a such that its square is even. You are proposing that a^2 is even, but a is odd. Using the definition of an odd number, define a and a^2.

c)   Determine what kind of a number a^2 is. Explain where the contradiction lies.

Proof by counter-example, or disproof

In order to prove that a statement is not true, it is only necessary to find one example that contradicts the statement. For example, if you were asked to prove that all prime numbers are odd, you would need only to state that 2 is a prime number, and it is even. You have thereby disproved the statement that all prime numbers are odd.

Work with a partner. Use either a GDC or computer algebra systems (CAS) software to make a table of values of n, and the factors of the expression $x^n - 1$.

🏛 **MATHS THROUGH HISTORY**

In the mid-1900s, Russian mathematician Nikolai Chebotaryov made it his hobby to factorise the expression $x^n - 1$ for ever-increasing integer values of n. He spent many years factorising this expression on paper, since there were no computers available when he was working on this.

n	$x^n - 1$	factors of $x^n - 1$
1	$x - 1$	$x - 1$
2	$x^2 - 1$	$(x + 1)(x - 1)$
3	$x^3 - 1$	
4	$x^4 - 1$	

When you have factorised a considerable number of expressions, make a conjecture regarding the coefficients of the variables in the factorized expressions. Test your conjecture by using other values for n.

Try to find a counter-example to your conjecture.

 ATL SKILLS
Thinking
Practise flexible thinking—develop multiple opposing, contradictory and complementary arguments.

Activity 2   Visual proofs

If a proof is based on a diagram and meets all three criteria of being irrefutable, able to be generalized and convincing, can it still be a proof even though there are no statements? This is something for you to think about at the end of this topic. For now, look at a few examples and decide if you can convert a "visual proof" into a formal proof in which you can justify all the steps, using results that you accept or have previously proved to be true.

STEP 1   The island problem

Mr Robinson visits BZA island, which fascinates him because it has the shape of an equilateral triangle. He decides that he wants to build a hut on this island. He tells his wife, a mathematician, that he wants to position the hut so that the total of the distances from the hut to the sides of the triangle is a minimum. His wife immediately tells him that he can then build the hut anywhere!

Look at Mrs Robinson's drawings showing several regular triangles. Then explain why he can build the hut anywhere. **Justify** your reasoning in detail.

TIP

Use properties of equilateral triangles to justify why some triangles and lengths on the diagram are equal.

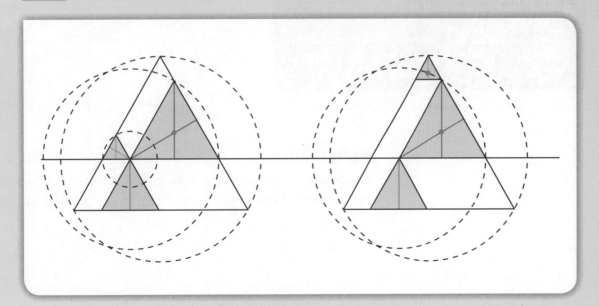

STEP 2 **Adding to infinity**

Study the diagram. Then explain how it proves the result shown. **State** any assumptions you make.

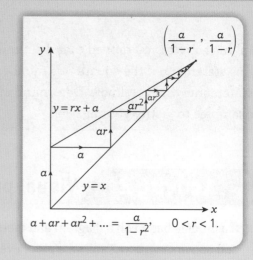

$$a + ar + ar^2 + \dots = \frac{a}{1-r}, \qquad 0 < r < 1.$$

STEP 3 **Circles and lines**

Consider two circles with diameters JL and KL. Assume this to be true, even if JL and KL do not look like diameters of the circles.

a) Explain how the diagram shows that H and I cannot be distinct points.

b) Use the diagram and part **a** to write an indirect proof of the conjecture "If two circles intersect at a point L and JL and KL are the diameters of these circles through L, then the line JK meets the circle at their other intersection point."

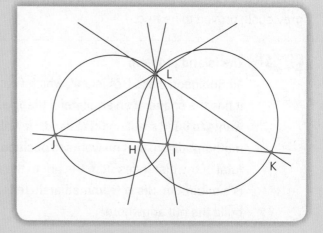

ATL SKILLS
Thinking

Practise flexible thinking—develop multiple opposing, contradictory and complementary arguments.

Reflection

After completing these activities, write a report about the validity of visual proofs. **State** clearly your position and **justify** your statements.

Empirical justifications in mathematics

People often use mathematics to draw conclusions and make predictions or informed guesses about real-world topics. To prove your point of view and convince others of its validity, you must be able to justify the mathematics and problem-solving techniques that you choose to solve such problems. Mathematics is often a very convincing tool because of its apparent criterion that something must be either right or wrong. In fact, it has also been used to justify statements that were later shown to be untrue. Being able not only to use mathematics in decision-making to justify a position, but also to analyse critically someone else's use of mathematics is a valuable skill.

🏛 MATHS THROUGH HISTORY

In 2008, the world experienced a financial crisis, the effects of which would be felt for at least a decade. This was mainly due to a mathematical formula called the "Gaussian copula function", which was formulated by Professor David X. Li, a brilliant contemporary mathematician. This formula modelled financial risk with such ease and accuracy that financial markets worldwide expanded and flourished to record levels. Over time, however, financial markets began to behave in ways not predicted by the model, revealing huge cracks in the model itself. Li himself said, "The most dangerous part is when people believe everything coming out of it [the model]". So using mathematics to justify real-world solutions requires great caution!

Predicting occurrences of natural disasters has always been a challenge for humankind. The survival and well-being of populations have often depended on their ability to identify patterns that allowed the prediction of catastrophic events such as hurricanes, tornadoes, forest fires, floods, volcanic eruptions, earthquakes or tsunamis. As history shows, humans have lost the battle a number of times. Cities have been destroyed and many lives have been lost, almost everywhere in the world. However, scientists have not given up and, with the help of technology, they have made progress in developing models that allow the prediction of some natural disasters such as hurricanes. For other phenomena the existence of a pattern for their occurrence remains undiscovered.

In the next activity you will analyse data about the eruptions of a famous volcano located in the south of Italy. Your goal is to identify a pattern for its eruptions and use statistical methods to justify your conclusions.

🔗 WEB LINKS

Go to http://www.history.com/topics and search for a video clip called "Deconstructing history: Pompeii". It will give you a sense of the magnitude of the eruption of Mount Vesuvius and its devastation of Pompeii in Italy.

This table shows the dates of the major eruptions of Mount Vesuvius since the one that destroyed the city of Pompeii in AD 79. Draw a graph to represent the data. Do not use a computer program. Then **use** your graph to answer the questions.

Eruption	Year
1	79
2	472
3	512
4	685
5	???
6	968
7	1037
8	1139
9	1500
10	1631
11	1660
12	1694
13	1698
14	1707

Eruption	Year
15	1737
16	1760
17	1767
18	1779
19	1794
20	1822
21	1834
22	1850
23	1855
24	1861
25	1868
26	1872
27	1906
28	1929

a) Determine the equation(s) of best fit. What model or models best describe(s) the pattern of the eruptions? If necessary, draw the curve in several parts. This is called a piecewise-defined function. You will need to find an equation for each part.

b) **Use** a GDC to check your result. **Justify** any choices you make. Explain any statistical and/ or algebraic reasoning that helps to **justify** how the data defines the domain of the piecewise function.

c) **Use** your equation to estimate the year of the fifth eruption. **Justify** your answer.

d) **Use** your equation to estimate the date of the next eruption after 1929. **Justify** your answer.

e) Research when the next eruption actually happened. Comment on the accuracy of your prediction in part d.

f) Based on your model, how many more eruptions should have occurred before today? Research the actual answer to this question and discuss any discrepancies.

g) Based on your model and research, is Vesuvius likely to erupt in your lifetime? Explain.

h) Does the fact that you can mathematically **justify** a model to fit a data set make it a valid choice? Explain.

GLOBAL CONTEXTS
Scientific and technical innovation

ATL SKILLS
Thinking
Interpret data.

INTERDISCIPLINARY LINKS

Individuals and societies

Apart from frequency of eruptions, research what other factors are important when analysing volcanic behaviour. The video clip you watched stated, "Experts believe another devastating Vesuvius eruption is imminent." Would you agree with this statement or is it dramatized for the video? Justify your answer.

TOPIC 3

Empirical justifications using algebraic methods

Throughout this chapter, you have justified your use of mathematics and your mathematical conclusions. However, mathematics is also used to justify all kinds of other decisions; sometimes it is just one part of the decision-making process. In this next task you will analyse the effect of an increased budget on the safety of a mission into space.

Activity 4 — Space shuttle safety

Shuttle missions into outer space are planned carefully so as to minimize the risk to both the craft and its occupants. Despite this, the Space Shuttle *Challenger* and the Space Shuttle *Columbia* missions both failed with many lives lost. Though the process is expensive, engineers do their best to foresee issues and add safety components to the shuttles to ensure a safe mission.

Suppose a shuttle's percentage chance of success (P) depends on the amount of money spent (d, in millions of dollars) on additional safety features. Suppose also that the function relating P and d is given by:

$$P(d) = \frac{300d}{3d+10}$$

In this scenario, if $P(d) = 0$, then the mission has a no chance of succeeding.

If $P(d) = 80$ then the mission has an 80% chance of succeeding and a 20% chance of failing.

a) Begin to sketch a graph of $P(d)$ by drawing the asymptotes first. Mark these clearly on your graph and include their equations.

b) Fill in this table. Make sure that these points are on your graph.

Amount spent in millions of dollars (d)	Percentage chance of success (P)
20	
40	
80	
100	
200	

c) What does the horizontal asymptote mean in real life?

d) Suppose you have been hired to determine how much money to spend on safety improvements for a shuttle mission. Make a recommendation showing how much you would spend on safety features. Explain the impact on the success rate. 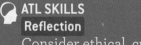 appropriate mathematics to justify your choice. In your decision-making process, you should also  the fact that the shuttle cost approximately $1.7 billion (without the improvements) to build. Write a short paragraph explaining your thinking and your mathematical justification.

🌐 **GLOBAL CONTEXTS**
Identities and relationships

👤 **ATL SKILLS**
Reflection
Consider ethical, cultural and environmental implications.

Reflection

What are the strengths and limitations of using algebraic and statistical methods to make predictions and empirical justifications in real-life situations?

Summary

In this chapter you looked at justifications by conducting formal and informal proofs and using visual models to conduct visual proofs. Through these examples you can see how important these proofs are in helping you develop your ability to communicate a logical line of reasoning and how these skills can be transferred to many other branches of mathematics. You also looked at empirical justifications, using algebraic and statistical methods. By working through examples you saw how using mathematics to make generalizations and predictions within a particular context can help in decision-making. This problem-solving process, along with your ability to use mathematics to justify your conclusions, are valuable lifelong skills.

Measurement

A method of determining quantity, capacity or dimension, using a defined unit

INQUIRY QUESTIONS

TOPIC 1 Making measurements
- How can you measure something?

TOPIC 2 Related measurements
- How can you prove what you discover through measurement?

TOPIC 3 Using measurements to determine inaccessible measures
- Is anything truly immeasurable?

SKILLS

ATL

✓ Apply existing knowledge to generate new ideas, products or processes.

✓ Consider multiple alternatives, including those that might be unlikely or impossible.

✓ Test generalizations and conclusions.

✓ Apply skills and knowledge in unfamiliar situations.

✓ Organize and depict information logically.

Number

✓ Solve problems in context, using ratios and proportions.

Algebra

✓ Solve rational equations.

Geometry and trigonometry

✓ Explore the properties of similar triangles.

✓ Determine lengths and heights of objects, using similar triangles and trigonometric ratios.

✓ Solve problems in context, using sine rule and cosine rule.

✓ Conduct geometric proofs.

✓ Discover the relationship between the lengths of tangents to a circle.

✓ Discover the relationships between segments in a circle.

✓ Explore the properties of cyclic quadrilaterals.

✓ Solve problems in context, using formulas for volumes of 3D geometric shapes.

OTHER RELATED CONCEPTS

Generalization   Representation   Justification

GLOSSARY

SI the modern metric system of measurement used worldwide in many fields.

Stadium the measure of a track where foot races were held. Because different places had different size tracks, a stadium measured different lengths to different people!

COMMAND TERMS

Explain give a detailed account including reasons or causes.

Justify give valid reasons or evidence to support an answer or conclusion.

Select choose from a list or group.

Use apply knowledge or rules to put theory into practice.

Introducing measurement

With agriculture and cooperation came a need for ancient people to be able to measure things such as grain, plots of land and even time. However, the first tools of measurement relied on body parts, such as the thumb, forearm and foot or on natural events such as the periods of the Sun and Moon. This led to communities measuring similar items with different units and even different devices. Regardless of this variation, there has always been a need and a desire to measure, including objects that are inaccessible. Some tribes of Native Americans who needed trees of a certain height in order to build longhouses or canoes would measure them in the following way.

They would walk to a position where they could bend over and see the top of the tree through their legs, which happens to give an angle (of elevation) of about 45°. What would they need to measure in order to determine the height of the tree?

However, even with modern technology, there are still measurements that are difficult to determine. In this chapter you will look at how to construct measuring devices, how measurements within shapes relate to one another and also how some measurements can be used to determine others that cannot be measured directly. Galileo would be proud!

TOPIC 1

Making measurements

To measure how long things are, how tall they are or how far apart they are, you use units of length. You compare the quantity you want to measure with a chosen unit. The most common units of length are the metre—the **SI** standard unit for length—and its submultiples, including the millimetre, centimetre or decimetre, and multiples such as the kilometre.

After you have chosen an appropriate unit, what do you do if you cannot measure an object directly? What measurements can you use to approximate its length, area or volume?

Activity 1   Investigating rulers

In ancient times people did not have the sort of rulers that you use. The idea was developed fairly recently— about a couple of thousand years ago. You are going to design your own ruler from first principles.

Suppose that you want to design a ruler for measuring various lengths, but with as few marks as possible. Investigate this problem for rulers of different lengths.

By marking off six unit lengths on a piece of wood, you can make a ruler to measure all the lengths from 1 to 6 units directly.

It is possible to achieve the same result with just two marks.

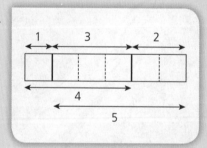

a) What is the minimum number of marks necessary to measure all the lengths from 1 to 12 units (whole numbers only)?

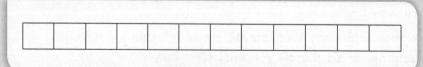

b) How many different solutions are there for a length 12 ruler?

c) What is the shortest ruler that requires three marks? Investigate rulers of different lengths.

d) What is the shortest ruler that requires four marks? Investigate rulers of different lengths.

e) Write down your conclusions. State, with reasons, whether a ruler with a minimal number of marks is a sensible choice for everyday use.

> **◯◯ WEB LINKS**
> Visit www.npl.co.uk, and go to "Educate + Explore>Factsheets> History of Length Measurement".

> **◯◯ INTERDISCIPLINARY LINKS**
> **Individuals and societies**
> The Ancient Greeks had no marks at all on their ruler! They called it a "straight edge" and the function of the ruler was simply to allow them to draw straight-line segments. To measure or compare lengths, the Greeks used another tool: a pair of compasses. With this tool they could easily construct a copy of a segment with a given length; they could add two lengths together and subtract a length from another one. The Greek ingenuity in operating with lengths, using just a straight edge and a pair of compasses, did not stop here. In fact, they discovered how to do all the arithmetic operations geometrically.

REFLECTION

Would it be possible to create rulers with minimum markings that could also measure half-units (such as 0.5, 1.5, 2.5, etc.)? **Explain** your reasoning.

> **🏛 MATHS THROUGH HISTORY**
> Many different units have been used, historically, to measure distances. The simplest were based on parts of the human body, but that brought obvious problems. The metric system came into use in the late 18th century and is now used in almost every country in the world.
>
> Rulers, similar in form to those available today, have been discovered at various archaeological sites. Excavations at Lothal, in the Indus valley, revealed a ruler dating from 2400 BC. It is calibrated to about $\frac{1}{16}$ inch or 1.6 millimetres.

GLOBAL CONTEXTS
Scientific and technical innovation

ATL SKILLS
Transfer
Apply skills and knowledge in unfamiliar situations.

You have seen that you can create all kinds of rulers to measure lengths. But how would you approximate the volume of an object when you have no suitable instrument to measure it? How were objects measured before rulers were designed? In the next activity you will attempt to find the volume of a hot air balloon by first choosing a suitable "ruler".

Activity 2   Up, up and away!

Every winter, balloon enthusiasts gather for a week of flying high over the Austrian Alps. They meet for take-off at Landhaus Koller in Gosau, Upper Austria.

Landhaus Koller in Gosau, Austria

© Family Koller

This balloon had its virgin flight in January 2011 during this competition.

© Family Koller

How could you approximate the volume of air in the balloons above?

Select one of the balloons. Find an appropriate object in the picture to use as a yardstick, or measuring tool. Then make a scale drawing of the balloon you have chosen.

You may use any models to find the volume, including geometric models: for example, spheres, cones, and so on. You may divide the shape into different geometric forms.

Justify your techniques and choice of model(s), and comment on the accuracy of your results.

Conduct research to determine how close your answer is to the actual volume.

a) How would your choice of yardstick affect the volume you calculated for the hot-air balloon? Give at least two specific examples.

b) Describe at least two other ways to improve the accuracy of your result. Be specific.

c) How significant is the effect of a small error (such as 10 cm) in the height of your yardstick on the overall estimate of the volume of the balloon?

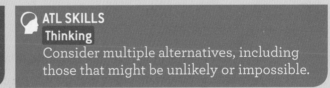

🌐 **GLOBAL CONTEXTS**
Scientific and technical innovation

💭 **ATL SKILLS**
Thinking
Consider multiple alternatives, including those that might be unlikely or impossible.

Related measurements

A total solar eclipse occurs when the Moon completely blocks the Sun's rays, as shown in the figure below.

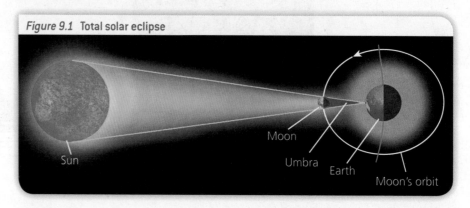

Figure 9.1 Total solar eclipse

Sun

Moon

Umbra

Earth

Moon's orbit

Only a small portion of Earth's population will see a total solar eclipse at any one time. They would all have to travel to the area indicated by the point of the triangle touching the Earth's surface. The geometry behind such an event demonstrates interesting properties of circles that you will investigate in the next activity. Is it possible to make a conjecture about relationships, based on measurements? Can you then justify or prove mathematically what you have found?

 Activity 3 **Tangents to circles**

STEP 1 **Use** a ruler or dynamic software to find the lengths of the tangents to each circle from the given point.

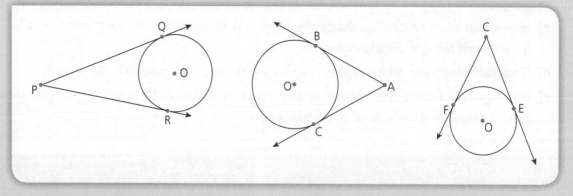

a) How are the lengths of the tangents related in each circle? Write the relationship in words.

b) **Use** geometry to prove that your conjecture is true. Hint: You know that a tangent and a radius are perpendicular to each other at the point of contact.

c) Now look again at the figure of the eclipse at the beginning of this topic. Determine which distances are equivalent and justify your answers.

STEP 2 On average, a total eclipse occurs somewhere on Earth approximately every 18 months, but only at the same location on the Earth once every 375 years. Using the relationships between measures, determine why a total solar eclipse is so rare.

a) Sketch a diagram to model this and show, by drawing tangent lines, how this occurs (it does not have to be to scale).

b) Using your knowledge of congruency and similar triangles, determine the relationships between the distances between the Earth, the Moon and the Sun. Formulate an equation to represent these relationships.

c) Given that the Sun is approximately 149 600 000 km from the Earth, the radius of the Moon is approximately 1740 km and the radius of the Sun is approximately 695 500 km, determine the furthest the Moon can be from the Earth for a total solar eclipse to occur.

d) Based on your results, explain why seeing a total eclipse of the Sun from your location is so rare. Give several reasons.

GLOBAL CONTEXTS
Scientific and technical innovation

ATL SKILLS
Thinking
Test generalizations and conclusions.

You already know that there is a relationship between the radius of a circle and its diameter. Did you know that there are also interesting relationships between the lengths of segments of chords of circles?

 Activity 4   **Lengths of segments of chords**

Look carefully at these diagrams. In each one, O is the centre of the circle and AB, CD are chords that intersect at M. The diagrams are not drawn to scale. Find the relationship between segments MA, MB, MC and MD in these two circles. Make a conjecture and test it. Then draw your own circle with two intersecting chords in it. Measure the segments and test your conjecture. Is it true for these lengths as well?

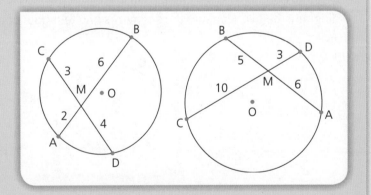

a) How are the lengths of the segments of chords in a circle related to each other? Write the relationship in words.

b) [Use] geometry to prove that your conjecture is true.

ATL SKILLS
Thinking
Test generalizations and conclusions.

You have found a relationship between the lengths of segments of chords that intersect inside a circle. What happens if they intersect outside the circle?

Activity 5   Points external to the circle

Look carefully at these diagrams. In each one, the chords are extended to intersect outside the circle. The diagrams are not drawn to scale. Find a relationship between the lengths of the segments in the first two circles. Then draw your own circle with two chords that intersect outside of it. Measure the segments and test your conjecture. Is it true for these lengths as well?

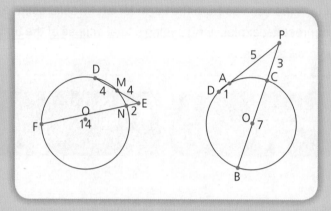

a) Find a relationship between the lengths of the line segments. Write the relationship in words.

b) [Use] geometry to prove that your conjecture is true.

What happens if one of the chords rotates so that there is a one chord and a tangent? Is there still a relationship between the lengths of the segments? Try to find a relationship for the lengths of the segments in these two circles. Then draw your own circle with a tangent line and a chord as in the diagrams below. Measure the segments and test your conjecture. Is it true for these lengths as well?

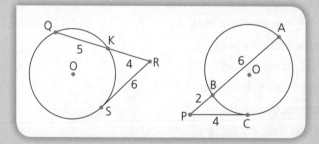

c) Find a relationship between the lengths of the line segments. Write the relationship in words.

d) [Use] geometry to prove that your conjecture is true.

You have investigated relationships among tangents, segments and chords in circles. In the next activity, you will investigate relationships among other measures involving circles and quadrilaterals.

 Activity 6   **Cyclic quadrilaterals**

A cyclic quadrilateral is a four-sided polygon with vertices that all lie on the same circle. As with any quadrilateral, you know that the sum of the interior angles is 360 degrees. But is there another relationship between the interior angles?

Use a protractor to measure the interior angles of each of these cyclic quadrilaterals. See if you can establish a relationship between them.

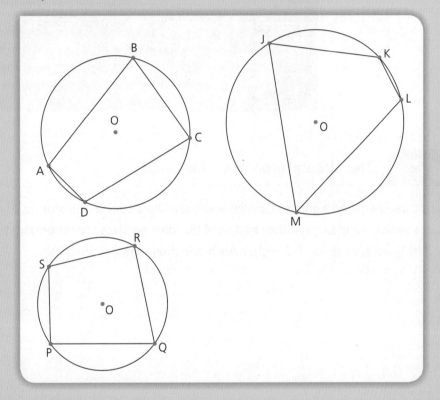

a) Find a relationship between the interior angles of a cyclic quadrilateral. Write the relationship in words.

b) **Use** geometry to prove that your conjecture is true.

REFLECTION

a) How are the rules you discovered in Activities 4 and 5 similar?

b) Do you think the rules you discovered were first found by measurement or by geometric proof? Why?

 ATL SKILLS
Thinking
Test generalizations and conclusions.

Using measurements to determine inaccessible measures

In this topic, you will explore an experiment conducted over 2200 years ago by a man named Eratosthenes.

> **🏛 MATHS THROUGH HISTORY**
>
> Eratosthenes was a mathematician, astronomer, geographer and poet who lived from 276 BC to 194 BC. He was the third librarian of the Great Library at Alexandria, which was widely considered the centre of science and learning in the ancient world. He is often credited with having coined the term "geography", using a system of latitude and longitude, and calculating the distance to the Sun as well as possibly even inventing the "leap year".

 Activity 7   The circumference of the Earth

Eratosthenes used a device called a gnomon, which was basically a stick in the ground. At noon on the day of the summer solstice, he measured the shadow of the stick and used trigonometry to calculate the angle of the Sun. If he were alive today, he might have made these measurements.

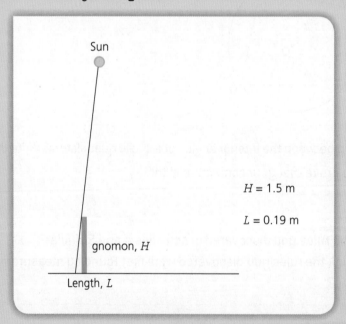

a) Find the angle of elevation of the Sun that Eratsothenes would have calculated.

b) Based on your result in **a**, determine the size of the third angle of the triangle.

Eratosthenes knew that the Sun was directly overhead at noon in Syene on that same day, since it shone down to the bottom of a deep well in that city. Therefore, Earth couldn't be flat: otherwise the Sun would have been directly overhead in Alexandria, too. Assuming that the rays of the Sun shine in a parallel fashion on Earth leads to the diagram below.

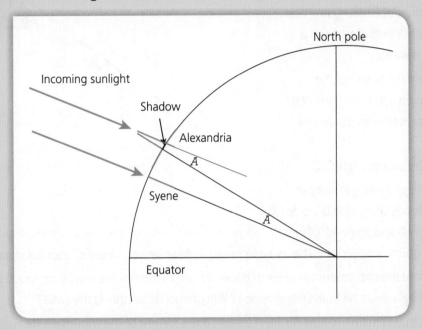

a) Eratosthenes' method

(i) **Justify** why the angle at the centre of Earth is also equal to A.

(ii) Determine the fraction of a circle that this angle represents.

Someone had measured the distance between Syene and Alexandria to be 5040 stadia (plural of *stadium*).

(iii) **Use** these facts to find the circumference of Earth measured in stadia.

(iv) There is great debate over the length of a **stadium**. An Attic stadium measures 185 metres, while an Egyptian stadium measures 157.5 metres. **Use** both of these to determine the circumference of Earth measured in kilometres.

How do your results compare to the actual circumference of the Earth, which is 40 075 km?
To what would you attribute any difference?

b) Biruni's method

Before looking at how Biruni calculated the circumference of Earth, do a little research.

(i) Who was Biruni?

(ii) When and where did he live?

(iii) What measuring tool(s) did he use?

> Here is another method for the determination of the circumference of the Earth. It does not require walking in deserts.
>
> Abu Rayhan Biruni

From the top of a mountain, Biruni measured the angle of depression or dip angle (α indicated in the diagram) to be 0.6 degrees.

Determine the radius of Earth, given that he was standing on a mountain with a height of 347.64 metres.

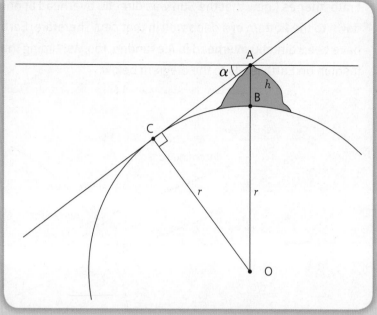

(i) Indicate the height of the mountain in the diagram. Write an expression for the length of OA.

(ii) Find the size of angle CAO.

(iii) **Use** trigonometry to write an equation relating side OA, side OC, angle CAO and angle ACO. Then solve your equation for r. The variable r may appear several times in your equation.

(iv) Calculate the circumference of Earth based on Biruni's measurement. How much closer to the actual figure was he than Eratosthenes? Why do you think this is the case?

The one question that remains is how Biruni measured the height of the mountain he was on. You will explore different methods for doing this in the summative task at the end of the chapter.

REFLECTION   **Is the Earth round?**

a) Aside from Eratosthenes' use of shadows, what evidence did people use to justify their belief that the Earth is round? What proof did they have or see?

b) Is the Earth actually round? If not, what shape is it? Why do we say that its circumference is 40 075 km?

🌐 **GLOBAL CONTEXTS**
Orientation in space and time

🧠 **ATL SKILLS**
Transfer
Apply skills and knowledge in unfamiliar situations.

🏛 **MATHS THROUGH HISTORY**
When Columbus sailed from Spain to find Asia, he used a circumference for the Earth that had been recently calculated by Arab astronomers. Unfortunately, they were wrong by about 30 per cent! Had he used Eratosthenes' value, calculated about 1600 years earlier, he would have known he had reached a New World, not his original destination, when he landed at Hispaniola in 1492.

WEB LINKS

Measuring the circumference of the Earth

There are a variety of YouTube videos that demonstrate both Eratosthenes' and Biruni's methods. One video even shows how someone recreated the same measurements as Biruni. Search YouTube for "Calculating the circumference of the earth by Al Biruni in Middle Ages (BBC)".

In the previous activity, you followed someone else's method for determining the measure of an inaccessible distance, the radius of the Earth. In the next activity, you will find the height of an object that is difficult to measure directly. You will use a variety of methods, one of which you will find for yourself.

Activity 8   Measuring the immeasurable

INTERDISCIPLINARY LINKS

Design

As a design task, research the problem of measuring angles of elevation and depression. Then, using everyday items, design your own clinometer to measures these angles.

In this activity, you will use a clinometer and trigonometry to calculate a height that can't be measured easily. You will use three different methods to calculate the height of an object and then reflect on how accurate you think your answers are and the usefulness of the methods.

The object

Find an interesting object with a height that would be difficult to measure directly. It is important that the object has some flat, level ground around it and that you can find a point directly under its maximum height. For example, choose a tree, a building or a flagpole. A mountain would not be a good choice since you can't stand directly under its highest point. It is very important that the object is not on a sloping surface. The points from which you take measurements must be on the same level as the base of the object.

Method 1

a) Standing away from the object, use a clinometer to measure the angle of elevation to the top of the object. Then measure the angle of depression to the base of the object. Finally, measure how far you are away from a point directly under the point where the object's height is a maximum (not necessarily the edge of the object).

b) Draw a diagram to represent the object. Show the place where you took the measurements and indicate the measurements on the diagram.

c) Use trigonometry to calculate the height of the object.

Method 2

a) Use the clinometer to measure the angle of elevation to the top of the object from a point away from its base.

b) Move a short distance towards the object and measure the new angle of elevation. Measure the distance between the two places where you took your measurements.

c) Draw a diagram to represent the object, the places where you took the measurements and the measurements themselves. Use trigonometry to calculate the height of the object. Remember to add the height at which you held the clinometer to your result.

Method 3

a) Use any other method of your choice to calculate the height of the object. You may research other methods but the one you select must require you to make measurements and use mathematics to find the height. You must not measure the height of the object directly.

b) Draw a diagram to represent the object, the places where you took the measurements and the measurements themselves. Use appropriate mathematics to calculate the height of the object.

REFLECTION

a) Do your answers make sense? Explain.

b) Discuss the accuracy of your results. Did all methods give similar answers? How close do you think you are to the actual height of the object? Compare your results to the real answer if you have access to that information.

c) Which method would be the most appropriate to measure the height of the mountain Biruni was on, or even of Mount Everest? Explain.

d) Research the Great Trigonometrical Survey of India to see which method was used and why it was necessary.

GLOBAL CONTEXTS
Scientific and technical innovation

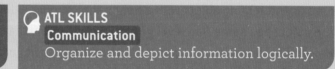
ATL SKILLS
Communication
Organize and depict information logically.

Summary

In this chapter, you started by looking at how you make measurements and how different types of rulers can be designed. Based on a few simple measurements, you found out that it is possible to approximate other measurements, such as volume or height. Being able to measure was also the beginning of being able to state conjectures related to measurements in a circle. Your conjectures could then be proved, using theorems in geometry that you either knew or developed. In the end, being able to measure and work with measurements is one of the first steps in discovering even more mathematics.

Model

A depiction of a real-life event using expressions, equations or graphs

INQUIRY QUESTIONS

TOPIC 1 Linear and quadratic functions
- ◼ **What is a mathematical model?**

TOPIC 2 Regression models
- ◼ **How reliable are mathematical models?**

TOPIC 3 Scale models
- ◼ **Is any model perfect?**

SKILLS

ATL

- ✓ Apply existing knowledge to generate new ideas, products or processes.
- ✓ Listen actively to other perspectives and ideas.
- ✓ Identify trends and forecast possibilities.
- ✓ Make unexpected or unusual connections between objects and/or ideas.
- ✓ Collect and analyse data to identify solutions and make informed decisions.
- ✓ Use models and simulations to explore complex systems and issues.
- ✓ Structure information in summaries, essays and reports.

Algebra

- ✓ Find the equation of a line, given two points on the line.
- ✓ Find the equations of linear and quadratic functions, given the graph.
- ✓ Define a quadratic function in standard and vertex forms.
- ✓ Define an absolute value function in vertex form.
- ✓ Describe a piecewise-defined function.

Geometry and trigonometry

- ✓ Find the distance between two points, the shortest distance from a point to a line and the areas of triangles, circles and sectors.
- ✓ Select the most appropriate triangle centre and justify why it is the best choice in a given context.

Statistics and probability

- ✓ Find a line of best fit by sight, and using technology.
- ✓ Use quadratic and exponential regression to determine an equation to model a real-life situation.

OTHER RELATED CONCEPTS

Space   Representation   Simplification   Justification

GLOSSARY

Correlation coefficient, r
a statistical measure of the strength of the relationship between two random variables.

Line of regression, line of best fit
a line that represents plotted points of observable data; the equation of the line models the relationship between the two variables.

COMMAND TERMS

Compare give an account of the similarities between two (or more) items or situations, referring to both (all) of them throughout.

Describe give a detailed account or picture of a situation, event, pattern or process.

Explain give a detailed account including reasons or causes.

Justify give valid reasons or evidence to support an answer or conclusion.

Select choose from a list or group.

State give a specific name, value or other brief answer without explanation or calculation.

Introducing model

You have already solved many mathematics problems in real-world contexts. There are mathematicians who are paid to find mathematical solutions to important real-life problems, such as global warming, population density, the spread of disease, and the causes and effects of an economic crisis. The process these mathematicians use is called mathematical modelling. In mathematical modelling, mathematics is used to:

- describe real-life situations
- explain real-life phenomena
- test possible solutions to real-life problems
- make predictions about real-life situations.

Just as an architect might make a model of a building to help understand the real building better without actually constructing it, mathematicians use mathematical models to help them understand a real situation without it actually happening. In a mathematical model you will use mathematical terms to simplify or simulate a real situation so you can explore its properties.

For any real-life problem, there may be several possible models that would fit the situation. The mathematician must investigate the merits of each model and choose the one that fits best.

Figure 10.1 explains the process of mathematical modelling.

First, you must clearly understand the real-life problem. Then, you must identify the variables, parameters and constraints or limitations of the problem.

Next, you can use different branches of mathematics (for example, algebra, geometry or trigonometry) to translate the problem into a model that could represent it. Finally, you can use appropriate technology to analyse your model and reach model solutions. When you have found a model solution, you will need to interpret it, in light of the problem, to decide whether the solution makes sense in the real-life context. Finally, you can modify your model to find solutions to other similar real-life problems.

The activities in this chapter will introduce different kinds of problems in mathematical modelling.

> *What distinguishes a mathematical model from, say, a poem, a song, a portrait or any other kind of "model", is that the mathematical model is an image or picture of reality painted with logical symbols instead of with words, sounds or watercolours.*
>
> John Casti

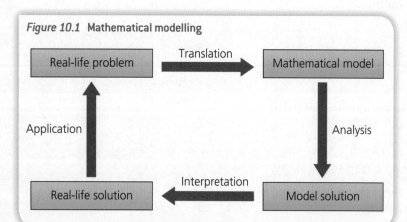

Figure 10.1 **Mathematical modelling**

Linear and quadratic functions

Linear and quadratic functions are probably the most common mathematical models you have worked with so far. Despite being fairly simple models, they can describe not only relations such as the position and speed of a projectile but also more complex situations such as those mentioned below, where you have to combine both types of function and consider domain restrictions (the specific values of the independent variable in the function you need to consider when graphing it).

 Activity 1   **Images**

In the production of animation films, an entire team of mathematicians works together to design the mathematical models that define the objects you see on the screen.

How do they create the functions that define the objects in the animations? Look at the static images below and think about the functions that define them.

... the process is about creating this large human-driven system, which is essentially a program that is capable of generating the right images in the right sequence ... there's data and mathematics that's being processed at a massive scale across thousands of individual computers. The ability to harness algorithms, fluid dynamics, and mathematics, and put that capability in the hands of someone who can express him or herself is the challenge.

Lincoln Wallen, DreamWorks Animation

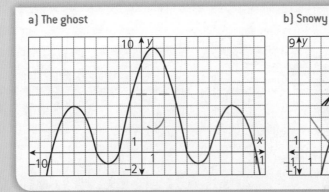

a) The ghost

b) Snowy

STEP 1   Write down the functions that you think define the shapes in the images above. Make sure you write down the domain of each function. Try to use as few functions as possible to define the image. Use the drawing feature of your GDC or software to draw the circle on top of Snowy's hat.

STEP 2   Use graphing software or a GDC to graph your functions and confirm that you obtain the images.

You have studied transformations of functions, such as horizontal and vertical shifts, reflections and dilations. Look at your list of functions and decide if you can modify any of them by using an appropriate transformation. **Describe** the transformation(s) necessary to obtain one function from the other.

 CHAPTER LINKS
See Chapter 2 on form to review different types of quadratic function.

 GLOBAL CONTEXTS
Personal and cultural expression

ATL SKILLS
Thinking
Apply existing knowledge to generate new ideas, products or processes.

Activity 2   Tunnel shapes

Technological progress has made tunnel construction more efficient and easier than it has ever been before. Despite some controversy about positive and negative impacts on the environment, tunnels are increasingly being built to shorten travelling distances between locations and increase the speed of transporting people and goods, both by road and train.

Tunnels do not all look the same. In this activity, you will start by investigating tunnels that have parabolic cross-sections. Then you will research and learn more about different types of tunnels, and why they have particular shapes.

The diagram shows the entrance to a parabolic tunnel. The maximum height of this tunnel is 4 metres and the maximum width is 6 metres.

STEP 1   **a)** By considering an appropriate set of axes, **determine** an equation of the curve that represents the parabolic tunnel. Give reasons for your choice of axes.

     b) Use graphing software or the GDC to draw the graph of your model accurately.

STEP 2 **a)** Let P be any point on the curve. Draw a rectangle, R, as shown in the diagram, with one side on the horizontal x-axis, and let P be one of the vertices of the rectangle.

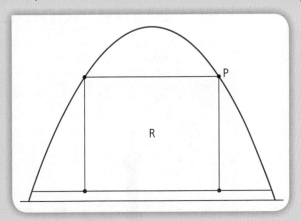

b) Now let a be the x-coordinate of P. Label the y-coordinate in terms of a, then find the area of R in terms of a.

c) Find the value of the maximum area of R. [Explain] how you have found this maximum area.

d) [Explain] why the area of this rectangle alone is not sufficient to set size limitations for vehicles going through the tunnel. [Describe] the factors that you need to consider in order to set such size limitations. Work out some possible sizes of trucks that can go through the tunnel. Make sure that you consider other aspects, such as safety.

STEP 3 **a)** Consider the cross-section of this tunnel. Estimate its area. [Explain] your method in detail.

b) Now write down the ratio between the area of the rectangle and the area of the cross-section of the tunnel.

STEP 4 [Explain] how you can use the ratio you calculated in step 3 as an indicator of the size limitations on vehicles that can enter the tunnel. [Explain] also how this ratio relates to other factors you discussed in step 2.

REFLECTION

a) [Describe] how your model may differ from real life.

b) [Describe] what other models you could have used in Step 3a. [Explain] how you decide which model is the best one.

c) Research and discuss advantages and disadvantages of tunnels with different shapes. Consider different factors such as cost, safety and environment.

d) [Explain] why modelling the shape of a tunnel is important for architects and planners.

GLOBAL CONTEXTS
Scientific and technical innovation

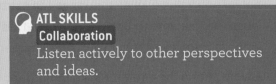
ATL SKILLS
Collaboration
Listen actively to other perspectives and ideas.

Regression models

Figure 10.2 Shanghai skyline

 INTERDISCIPLINARY LINKS

Design

If you look around you at buildings and other structures, you can easily find mathematical shapes you have studied. In cities such as Shanghai, cuboids, spheres, cylinders and combinations of other well-known solids populate the skyline. Many factors influence the choice of the shape of a building, from its geometrical properties that offer, for example, greater stability to simply its aesthetic appeal.

In this topic you will use the **correlation coefficient** r, from your GDC, for linear models. The correlation coefficient r measures both the strength and direction of a linear relationship between two random variables. Its absolute value varies between 0 and 1, $|r| \leq 1$, with values closer to 1 indicating a stronger relationship and values closer to 0 indicating a weaker relationship. The sign of r indicates the direction of the relationship, or the gradient of the linear model. A positive value indicates that as the values of one variable increase so do the values of the other variable, whereas a negative value indicates that as the values of one variable increase the values of the other variable decrease.

Activity 3 — Going for gold

Linear regression is a modelling technique that allows you to make predictions about the value of one variable:

- when you know the value of another variable
- when the value is within the given interval; for example, if a variable is defined for values between 2 and 10, you can make a valid prediction for any value within the interval (an interpolation) but not for any value outside this range (an extrapolation)
- when the relationship between the variables can be modelled by a linear function. Drawing a graph is a good first step in deciding whether or not a pattern could be linear.

The data table shows the times for the Olympic gold medallists in the men's and women's 100 m race. Use the data to draw a graph and answer the questions that follow.

Winning times of the women's and men's 100 m race at the Olympics					
Year	Women's winning time (seconds)	Men's winning time (seconds)	Year	Women's winning time (seconds)	Men's winning time (seconds)
1928	12.2	10.8	1976	11.08	10.06
1932	11.9	10.3	1980	11.06	10.25
1936	11.5	10.3	1984	10.97	9.99
1948	11.9	10.3	1988	10.54	9.92
1952	11.5	10.4	1992	10.82	9.96
1956	11.5	10.5	1996	10.94	9.84
1960	11.0	10.2	2000	11.12	9.87
1964	11.4	10.0	2004	10.93	9.85
1968	11.0	9.95	2008	10.78	9.69
1972	11.07	10.14	2012	10.75	9.63

STEP 1 Draw up two coordinate grids. Plot the women's winning time against the year on one grid and the men's winning time against the year on the other.

STEP 2 Draw the **line of best fit** for each graph and explain what the gradient means in this context. The line of best fit is also known as the **line of regression**.

INTERNATIONAL MATHEMATICS
In certain parts of the world you might use the word *slope* instead of *gradient*.

STEP 3 Find the equation of the line of best fit for each graph.

STEP 4 Some years have no data. Use your model to predict what the winning times would have been. Is it appropriate to assume that your model applies for these years? Explain.

STEP 5 Based on your model, what times would you predict for the winners of the women's and men's 100 m races at the Rio de Janeiro Olympics in 2016? Justify your answer by showing your work.

STEP 6 According to your models, will women ever run the 100 m as fast as men? How do you know? If so, find the year when this will happen and the winning time of the race.

REFLECTION

a) Conduct research to find out why there is no data for some of the years.

b) How do you think that the graphs for future years will change? Is it appropriate to assume that your models will accurately represent the winning race times forever? Explain. If not, what model(s) would be most appropriate to describe any new patterns?

c) If you used your regression equation to predict a value in the table, will your answer necessarily be the one in the table? **Explain** your reasoning.

QUICK THINK

Does the scale really matter? The choice of scale for graphing data is very important, as it influences the conclusions that are made about the information being displayed. Consider these two graphs, both of which show the increase in price of a particular item over a three-year period.

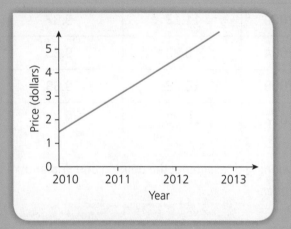

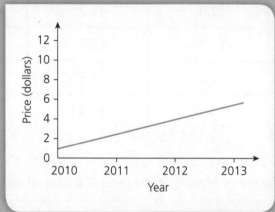

Explain what you think the data is meant to convey, and how the different scales affect the message.

👥 **Activity 4**   **Dental arches**

STEP 1   If you have overcrowded or misaligned teeth, a dentist or orthodontist may take an impression of your mouth to study your dental arch. Models of teeth are also used in sports dentistry to create teeth and mouth guards for use in contact sports.

a) Take a piece of paper and fold it in half. Make sure you can fit the paper into your mouth. Push it as far back as you can, then bite down on it. This will give you an impression of your mouth.

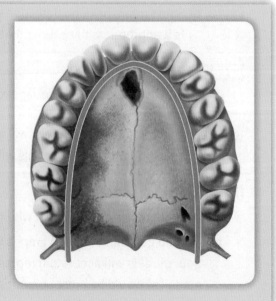

b) Use a dark marker to mark each tooth indentation clearly, with a point.

    (i) You will need to mark at least 10 points of your dental arch.

    (ii) Two of your points should represent the ends of the dental arch (the back teeth) and one (or two) should approximate the maximum height of the arch (the front teeth).

c) Trace the points of your bite onto a sheet of graph paper.

d) Draw a set of axes over your points so that the approximate maximum value of your dental arch lies at the origin.

e) Identify and label the coordinates of each point.

f) Based on your points, find an equation to model your dental arch. Show all the steps in your work and include the points you have chosen. **Explain** why you selected them.

g) Now enter the points into a GDC or use graphing software to identify the parabola of best fit. By researching the meaning of r and r^2 (the coefficient of determination) for non-linear models, determine how well the parabola fits as a model of your dental arch. **Justify** your reasoning.

h) **Compare** the two equations you found in questions **f** and **g** and **explain** why they may be different.

STEP 2 Custom-made mouth guards may be expensive. Some companies sell standard guards that you can adapt to fit your teeth. These companies need to make sure that the guards they produce will fit as many people as possible, so they collect data about as many mouth and teeth shapes as they can.

a) Collect tooth indent data from other people in the class. How many other sets of data do you think you need to produce an accurate model?

b) Use all of your data to find an equation to represent a standard dental arch.

c) Draw the parabola from your equation. How well does it model the standard dental arch? Would you expect it to be a better or worse fit than the parabola that fits your individual dental arch?

d) **Compare** the standard dental arch and your individual dental arch and **explain** why they may be different.

e) The company needs the average width and height of a dental arch to make a standard guard. Calculate these lengths, showing all the steps in your work.

REFLECTION

One type of generic mouth guard is made from plastic. It is wide and long, enabling you to cut the ends off so that it doesn't extend past your back teeth. To shape it to your mouth, you put it in hot water, which softens the plastic so that it can be shaped to the teeth. This will also allow it to shrink, to fit almost any mouth. When creating a mouth guard like this one, would you alter the equation of your regression parabola at all? If so, what changes would you make?

GLOBAL CONTEXTS
Scientific and technical innovation

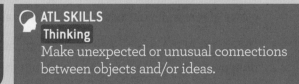
ATL SKILLS
Thinking
Make unexpected or unusual connections between objects and/or ideas.

Why are there fluorescent mosquitoes in Malaysia, northern Brazil, the Cayman Islands and possibly the United States? It's not an invasion but rather part of a defence against dengue fever. Dengue fever is a flu-like illness that is spread by mosquitoes and affects humans, causing a variety of symptoms. It has also been called *breakbone fever*. Although there are some treatment methods, there is no vaccine and no effective cure.

With numbers of cases growing rapidly and over 40 per cent of the world's population at risk from infection, the World Health Organization (WHO) issued a warning in 2013 on the potential for dengue fever to become a global health hazard.

STEP 1 The data below gives some approximate numbers relating to dengue fever infections.

Year	Dengue fever cases reported to the WHO
1957	900
1965	15 500
1975	122 000
1985	296 000
1995	480 000
2005	970 000

a) Display the data on a graph. Choose an appropriate scale to represent the data accurately.

b) **State**, with reasons, the type of function you think best describes the data.

c) Use a GDC or graphing software to experiment with different types of function. **Select** the one that you think best models the data. **Justify** your choice.

d) Use your model to find the number of dengue fever cases reported to the WHO this year.

e) Based on your model, determine when the number of cases will reach one billion.

STEP 2 The rapid spread of dengue fever is a real cause for concern. Researchers seeking a solution have found a method to combat the dengue-carrying mosquitoes. This is where the fluorescent mosquitoes come in.

Dengue fever is passed to humans through the bite of a female fever-carrying mosquito. Male mosquitoes do not carry the disease. Scientists have found a way of combating dengue fever that seems to have worked in several countries where it has been tested. One of the features of the technique involves modifying some mosquitoes so that they carry a special protein that glows green under ultraviolet light. This enables them to be tracked and counted.

The males of the group have been genetically modified so that, in the wild, they can mate with dengue-carrying female mosquitoes, but their offspring will not survive to adulthood.

How did researchers know that their plan was working? They used ultraviolet light sources and counted the glowing mosquitoes. This data is an example of their results.

Time (weeks)	Number of glowing mosquitoes
0	1000
4	770
8	580
12	438
16	333

The numbers of glowing mosquitoes fell dramatically. Such a fall in mosquito numbers could reduce the spread of dengue fever in just a few generations.

a) Graph the data. Choose a reasonable scale.

b) **State**, with reasons, the type of function you think best describes the data.

c) Using a GDC or graphing software, experiment with different types of function. **Select** the one that you think best models the data. **Justify** your choice.

d) Based on your model, determine how long it will take **(i)** for only 10 per cent of the population to remain **(ii)** for only 1 per cent to remain.

REFLECTION

a) In real life, is it possible to eradicate the dengue fever-carrying mosquito population completely? Does your model allow for this possibility? **Explain** your answer.

b) If the genetically modified mosquitoes were used in an entire country, is it appropriate to assume that the same model would apply in that situation? **Explain** your answer.

c) **Explain** why people might be against introducing genetically modified mosquitoes into the environment.

d) Research another method that is being suggested to control the spread of dengue fever. Would the same type of model apply there? **Explain** your answer.

GLOBAL CONTEXTS
Globalization and sustainability

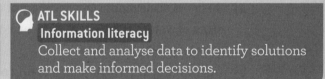

ATL SKILLS
Information literacy
Collect and analyse data to identify solutions and make informed decisions.

Scale models

A scale model is a representation of an object or objects that may be smaller or larger than the original yet keeps the same mathematical proportions. The ratio of the length on the model to the corresponding length on the original is called the *scale factor*. Scale models are used in many fields such as art, film-set design, engineering and architecture. They can enable you to design the best possible product/ solution, as you can test the validity of the design to ensure the optimal solution has been found at minimal expense. In the following activities you will create a 2D scale model of an irrigation system to ensure resources are not wasted, and an urban planning model to determine the optimal location of resources to meet the needs of the people that will be living within that community.

 Activity 6   **Sprinklers**

A crop field measures 85 m by 75 m.

You can use any quantity and combination of sprinklers. There are three types.

- Sprinkler one has a radial range (the radius of the circle covered by a sprinkler) of 25 m.
- Sprinkler two has a radial range of 20 m.
- Sprinkler three has a radial range of 15 m.

Pipe costs $40 per metre, fully installed. The sprinklers must be connected to a water pipe that runs along one of the widths of the field.

Determine the best way to configure the sprinklers and find the cost of installing the irrigation piping.

STEP 1   Find the optimal configuration.

a) Design a scale model to test different combinations and sizes of water sprinkler. You may use computer software to help you if you wish.

b) Try a variety of configurations to find out which one will yield the greatest area covered, with no overlap, while minimizing piping between sprinklers and staying within the crop field.

c) Draw a labelled 2D scale model of your configuration in the crop field. You must include the scale factor you used.

TIP

Your scale model will be an arial view of the entire crop field with the three types of sprinkler represented as circles showing their radial range within the rectangular field. Do this on paper or use dynamic geometry software so you can move the circles around and test the different sizes.

d) Write a detailed explanation of why your configuration is the best, including other sketches and noting why these other configurations are not as good as yours.

STEP 2 Determine the cost of irrigation piping.

a) Determine the percentage of cropland that will not be covered by the sprinklers.

b) Connect the sprinkler heads (the centre of each of the three circles where the water sprays out) to form an enclosed shape and determine all side lengths and angles. Label these clearly on your diagram.

c) Within the boundaries of the sprinklers (the shape created by connecting the sprinkler heads), is there a region of the crop field that will not be covered by the sprinklers? If so, determine its area.

d) Determine the piping cost to install the irrigation system, based on your configuration.

REFLECTION

a) Is your configuration the optimal one?  how you know.

b) Is your answer as accurate as it should be? the degree of accuracy you used.

c) What suggestions could you make about the sections of the field that will not be covered by the sprinklers?

d) **Explain** any other real-life factors that you would have to consider when setting up an irrigation system like this.

e) Suggest other situations in which you could use geometry and trigonometry to solve problems to help minimize human impact on the environment.

> **TIP**
>
> Use trigonometric and geometric formulas for triangles and sectors to calculate the area and use analytic geometry techniques to find the shortest length of piping required to minimize costs. The QUICK THINK that follows this activity could help you determine a formula to calculate the area of the triangle.

🌐 **GLOBAL CONTEXTS**
Fairness and development

🧠 **ATL SKILLS**
Thinking
Use models and simulations to explore complex systems and issues.

🔗 **CHAPTER LINKS**
See Chapter 14, Simplification for more on the relationship between sides of right-angled triangles.

> **TIP**
>
> You can use Heron's formula to find the area of any triangle if you know three sides.

QUICK THINK

Find the equation for the area of a triangle. Refer to triangle ABC, with lengths as shown.

Use trigonometry to express the height in terms of an angle and side.

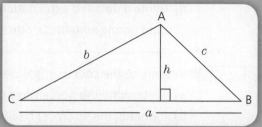

Then substitute your expression for the height into the basic formula for the area of a triangle. This will allow you to determine the equation for the area of a triangle, using only side lengths and an angle.

 Activity 7   Urban planning

You have been commissioned to design a residential community, in a currently uninhabited region, to support the needs of all of the people who will be living in three residential housing estates. The three separate housing estates are designed to be similar in size and density, and their centres are positioned on a map at the locations $(1, 1)$, $(2, 12)$ and $(20, 4)$. Each has a radius of approximately 0.5 kilometres. There is currently a train line that runs along the line $y = -2$. You have been given permission to add one station to that line. You will create a 2D scale model of your design for the residential community. Include all the infrastructure and services needed to support the community.

Use a scale of 1 cm : 0.5 km on the model.

What aspects do you have to think about when planning (or expanding) a residential community? What features will attract people to live in your town?

Your report must include:
- a one-page 2D scale model of your urban plan—you can use dynamic geometry software to help you
- explanation of all infrastructure (buildings and services) you have included in your residential community and why you have included it
- explanation of location of all of the infrastructure in reference to each other and the three housing estates and the set-up of major roads between them. **Explain** what mathematical concepts you used to determine the locations.

QUICK THINK

There are a few major cities around the world that have been planned from their inception. Research these cities and comment on how successful the city design has been in terms of livability.

TIP

As there are three housing estates, think about the different triangle centres which could be used for the location of different key infrastructure features.

- calculations of distances between key infrastructure features in reference to each other and the three housing estates
- discussion of other infrastructure (utilities and so on) that will need to be considered in the building process of the residential areas. No calculations are needed.

CHAPTER LINKS

See Chapter 15 on space for more on investigating triangle centres.

GLOBAL CONTEXTS
Globalization and sustainability

ATL SKILLS
Communication
Structure information in summaries, essays and reports.

🏛 **MATHS THROUGH HISTORY**
Hippodamos of Ancient Greece (498–408 BC) is considered by many to be the father of urban planning. In approximately 450 BC he arranged the buildings and the streets of Miletus to allow for the optimal flow of winds through the city and provide cooling during the hot summer. He was known for his grid plan, which he had developed on inspiration from geometrically designed settlements, which many Ancient Greek cities subsequently used.

Summary

You have been working on mathematical modelling activities covering a wide range of situations and mathematical topics. The models that you have created in these activities are not limited to the problems presented in this chapter, but rather they apply to a wide range of real-life situations. The next time you see an animated film, remember that hundreds of mathematicians may have worked behind the scenes to create your favourite characters and scenes. Mathematics plays a leading role in many different areas, including tackling serious world problems, fighting the spread of disease and making the best use of the Earth's finite resources.

Pattern

A set of numbers or objects that follow a specific order or rule

TOPIC 1 Famous number patterns
- **Why are some patterns more significant than others?**

TOPIC 2 Algebraic patterns
- **How does one pattern lead to another?**

TOPIC 3 Applying number patterns
- **Are patterns in nature an accident?**

SKILLS

ATL

✓ Understand and use mathematical notation.

✓ Propose and evaluate a variety of solutions.

✓ Consider multiple alternatives, including those that might be unlikely or impossible.

✓ Encourage others to contribute.

✓ Practise visible thinking strategies and techniques.

✓ Make connections between subject groups and disciplines.

✓ Consider ideas from multiple perspectives.

Number

✓ Calculate absolute value.

✓ Find the common factor of two integers.

Algebra

✓ Determine patterns in recursive functions.

✓ Find and justify general rules.

✓ Find and justify general rules for formulae.

✓ Use the quadratic formula and factorizing to solve quadratic equations.

✓ Factorize algebraic expressions.

✓ Determine the general rule and function that represents linear and quadratic patterns.

Geometry and trigonometry

✓ Use the properties of similar shapes to determine patterns.

OTHER
RELATED
CONCEPTS

Generalization   Representation   Justification   Model

GLOSSARY

Asymptote a line or curve to which the graph of a function gets arbitrarily close as the x-values or the y-values approach infinity.

General rule or term an expression that generates all of the elements of the pattern in study. In general, the rule can be determined explicitly as a function of the independent variable; sometimes however the rule may be described by an equation.

COMMAND TERMS

Describe give a detailed account or picture of a situation, event, pattern or process.

Explain give a detailed account including reasons or causes.

Solve obtain the answer(s) using algebraic and/or numerical and/or graphical methods.

State give a specific name, value or other brief answer without explanation or calculation.

Use apply knowledge or rules to put theory into practice.

Introducing pattern

A mathematician, like a painter or a poet, is a maker of patterns. If his patterns are more permanent than theirs, it is because they are made with ideas.

GH Hardy

From your earliest encounter with numbers and their properties to the study of geometric properties of shapes, from trigonometric ratios to properties of functions, from investigating patterns to arriving at conjectures to modelling real-life phenomena, you continually experience many distinct patterns. You may have also discovered some famous patterns in nature, such as the Fibonacci sequence, the golden ratio and Euler's number, the important natural constant e.

The search for patterns has always challenged mathematicians. The quest to find patterns for the occurrence of prime numbers or in the digits of pi has engaged various mathematicians through hundreds of years. Even today some of the "unsolved" problems in mathematics relate to unexplained patterns. Some institutions offer huge financial rewards for explanations or proofs for these unsolved problems.

Is there some pattern to chaotic phenomena? Chaos theory is among the newest branches of mathematics. The use of the word *chaos* might imply a lack of order, that things behave randomly. In fact, mathematicians are discovering that beneath the apparent random behaviour of dynamic systems such as global weather, the stock market and chemical reactions, as well as natural phenomena such as heartbeats, waterfalls, rapids, rock formations and snowflakes, there is enormous order. Chaos theory is helping mathematicians to understand and predict the behaviour of these systems.

In this chapter, you will explore some well-known number patterns and then look at ways of generating other numerical and geometrical patterns.

Figure 11.1 The study of the Lorenz attractor as part of chaos theory helps medical researchers understand heart attacks and predict them before they occur.

© Paul Bourke

Famous number patterns

Number patterns have always fascinated mathematicians and non-mathematicians. Some special patterns appear repeatedly in different situations and have gained a special place in the building of mathematics. Now it is your turn to rediscover some of these patterns and explore their applications.

 Activity 1 **Triangular numbers**

Use a piece of squared paper or graph paper.

Shade one square, in the top left-hand corner.

Next, shade the first two squares in the row below it.

Then shade the first three squares in the next row. Continue this process, shading an extra square each time, until you have completed eight rows.

Row, n	T_n – the total number of shaded squares
1	1
2	3
3	
4	
5	
6	
7	
8	

Can you predict, without actually drawing them all, how many squares will be shaded for each row of this pattern? To help you, take a different colour as illustrated, and complete the "staircase" to make a rectangle. Determine the area of the rectangle, and halve that number.

Can you now find a formula for T_n given any n?

T_n

CHAPTER LINKS

In Chapter 5, Change, there is a story about Gauss when he was in school. How does the pattern you discovered above relate to the process he went through in order to solve the question so quickly?

REFLECTION

Why do you think that the sequence of numbers formed with the entries in the 2nd column is called the *triangular numbers*?

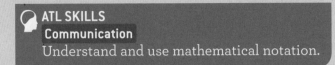
Polygonal numbers

A polygonal number p is a number that can be represented by n dots in the shape of a regular k–sided shape. The most well-known polygonal numbers are the square numbers 1, 4, 9, 16, ... and the triangular numbers 1, 3, 6, 10, ... but the process of obtaining polygonal numbers can be generalized for other regular k–sided shapes. For example, $k = 5$ produces the pentagonal numbers 1, 5, 12, 22, ... and $k = 6$ produces the hexagonal numbers 1, 6, 15, 28, ...

These numbers have many interesting properties that you can identify by observing their graphical representations, as shown in the diagrams below. For example, the triangular numbers start with one dot and then you add two dots, three dots, four dots, and so on, which leads to the recurrence formula $p_n = p_{n-1} + n$.

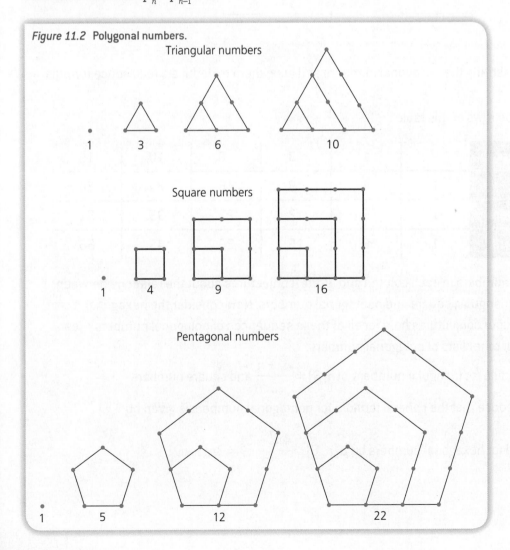

Figure 11.2 **Polygonal numbers.**

Triangular numbers

1 3 6 10

Square numbers

1 4 9 16

Pentagonal numbers

1 5 12 22

a) After you have read the discussion above, use a similar reasoning to deduce recurrence formulas for square and pentagonal numbers.

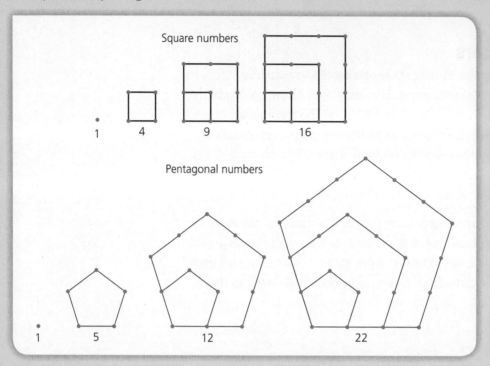

Square numbers

1   4   9   16

Pentagonal numbers

1   5   12   22

b) Draw diagrams representing the hexagonal numbers and use them to deduce a recurrence formula for this sequence.

c) Consider the first three rows of this table.

Triangular numbers		1	3	6	10	15
Square numbers	1	4	9	16	25	36
Pentagonal numbers	1	5	12	22	35	51
Hexagonal numbers	1	6	15	28	45	66

Look carefully at the numbers in each column and state conjectures about the relations between these sequences of triangular, square and pentagonal numbers. Now consider the hexagonal numbers as well. Do your conjectures hold for all of these sequences of polygonal numbers? Test the relation further for other lists of polygonal numbers.

d) Using the general formula for triangular numbers $p(n, 3) = \dfrac{n^2 + n}{2}$ and square numbers $p(n, 4) = \dfrac{2n^2}{2} = n^2$, deduce that the general formula for pentagonal numbers is given by $p(n, 5) = \dfrac{3n^2 - n}{2}$ and for hexagonal numbers by $p(n, 6) = \dfrac{4n^2 - 2n}{2} = 2n^2 - n$.

e) Observe the formulas for the k-polygonal numbers when $k = 3, 4, 5$ and 6. **Explain** any patterns you observe in these formulas. **State** your conjecture for the formula of $p(n, k)$. Test your formula further and write down your conclusion.

You have been looking at sequences of polygonal numbers. You will now look at how they can be used in an everyday situation.

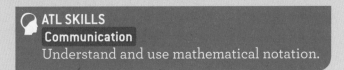

Activity 3   **Square and cubic routes**

STEP 1   Sarah lives in Square City, where all the blocks are squares. Every morning she has to walk from her building (A) to school (B).

This is a plan of her neighbourhood.

Work out all the possible shortest routes from her building to school. **Use** a grid to draw all possible routes.

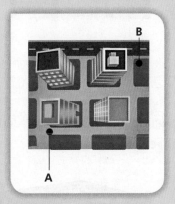

STEP 2   Billy also lives in Square City. This plan shows the location of his flat (A) and his school (B).

How many different "shortest" ways can Billy take from his flat to school?

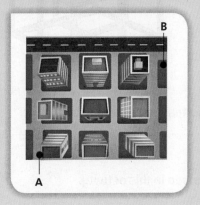

At each point in the grid, write the number of different shortest routes from A to that point. Try to find a pattern or a rule. **Explain** .

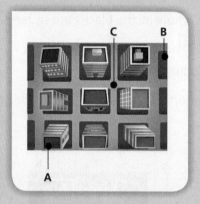

Tommy is a friend of Billy. Usually they go together to school. How many possible shortest routes to school can Billy take so that he meets his friend at the crossing C?

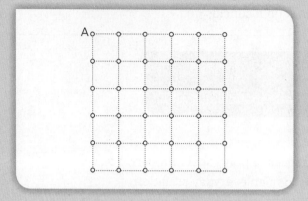

EXTENSION

Imagine now a cubic building under construction. Suppose that you want to go from A to B but you can only walk or climb along the edges of the cube.

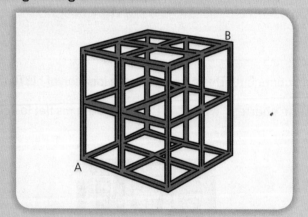

Work out the number of different shortest paths from A to B.

REFLECTION

Research the pattern you have observed in this activity.

GLOBAL CONTEXTS
Orientation in space and time

ATL SKILLS
Thinking
Propose and evaluate a variety of solutions.

INTERNATIONAL MATHEMATICS Pascal's triangle was known by other cultures long before Blaise Pascal, the French mathematician, first wrote about it in 1653. It appeared in an early Sanskrit work from the second century BC, and in Persian, Italian and Chinese works all dating to the 11th and 12th centuries.

You may be surprised to know that you have probably seen Pascal's Triangle (or a portion of it) before and you didn't recognize it by its name. The following activity will show you other instances in which it can occur or be found.

Activity 4   The binomial theorem

a) Use multiplication to expand these expressions. You can use answers from one part to help you with others.

(i) $(x+y)^2$ (ii) $(x+y)^3$ (iii) $(x+y)^4$

b) What patterns do you see in the exponents of successive terms in your answers? Why does this happen?

c) Begin with $(x+y)$ and write the coefficients of each expansion in its own row, to obtain this array.

$$
\begin{array}{ccccccc}
 & & & 1 & 1 & & \\
 & & 1 & 2 & 1 & & \\
 & 1 & 3 & 3 & 1 & & \\
1 & 4 & 6 & 4 & 1 & & \\
\end{array}
$$

What patterns do you notice in these coefficients? How would you find the next two rows? Write them down.

d) You should have noticed that this is Pascal's triangle, a famous pattern that was mentioned in the International Mathematics box just above this activity. What other patterns do you see in it?

$$
\begin{array}{ccccccccccc}
 & & & & & 1 & & & & & \\
 & & & & 1 & & 1 & & & & \\
 & & & 1 & & 2 & & 1 & & & \\
 & & 1 & & 3 & & 3 & & 1 & & \\
 & 1 & & 4 & & 6 & & 4 & & 1 & \\
1 & & 5 & & 10 & & 10 & & 5 & & 1 \\
\end{array}
$$

Based on your results, write down a polynomial expression equivalent to $(x+y)^6$ without multiplying.

What you have just discovered is the basis for the binomial theorem. The binomial theorem provides a formula that allows you to write down the expanded forms of powers of a binomial without having to carry on long calculations.

What happens if one of the terms in the expression is a number? What happens if the coefficients of the variables within the expression are not 1?

e) Use multiplication to expand each expression.

  (i) $(x+3)^2$ **(ii)** $(x+3)^3$

  (iii) $(x-3)^4$ **(iv)** $(2x+y)^3$

f) Which patterns from the previous exercises also occur here?

g) How is it possible to determine the coefficient of any term in the expanded form?

h) Describe , in writing, a shortcut for expanding binomials.

i) Use your shortcut to expand these expressions without multiplying.

  (i) $(2x+y)^3$ **(ii)** $(3x+2y)^4$ **(iii)** $(x-2y)^3$

j) **(i)** Without expanding $(2x-3y)^6$, determine the coefficient of the x^2y^4 term.

  (ii) Without expanding $(4x+2y)^7$, determine the coefficient of the x^5y^2 term.

  (iii) Without expanding $(x+3y)^5$, determine the coefficient of the x^2y^3 term.

  (iv) Without expanding $(5x-4y)^4$, determine the coefficient of the xy^3 term.

k) How would you use this method to approximate 1.1^{10} to 4 decimal places without actually multiplying it out or using a GDC?

TIP

You may want to write 1.1 as $1 + 0.1$, so this becomes $(1 + 0.1)^{10}$. In a similar way, approximate 1.96^{10} to 4 decimal places.

REFLECTION

a) Did Pascal's triangle emerge from the binomial theorem or is the binomial theorem an application of Pascal's triangle? Explain your reasoning.

b) If you can always work out the correct expression by multiplying, why do you need to learn the binomial theorem? When is it useful?

ATL SKILLS
Thinking
Propose and evaluate a variety of solutions.

STEP 1   How can you solve equations that contain an endless pattern?

a) **Solve** the endless equation below.

$$x=\sqrt{1+\sqrt{1+\sqrt{1+\sqrt{1+\cdots}}}}$$

TIP
Hint: Can you simplify the right-hand side?

Notice that the expression after the "1 +" under the square root sign is the same as x and suddenly what appeared endlessly difficult has been made simpler!

You now have $x=\sqrt{1+\sqrt{1+\sqrt{1+\sqrt{1+\cdots}}}}=\sqrt{1+x}$

which reduces to

$$x=\sqrt{1+x}$$

Squaring both sides:

$$x^2=1+x$$

Use the quadratic formula to **solve** the equation. Make sure that you check your solutions in the original equation, $x=\sqrt{1+x}$, for any extraneous solutions.

b) By using a similar approach as in **a**, choose an appropriate substitution for the right-hand side and **solve** :

$$x=1+\cfrac{1}{1+\cfrac{1}{1+\cfrac{1}{1+\cfrac{1}{1+\cdots}}}}$$

Check your answers in the original equation. Why must you apply the condition $x>1$?

c) Look at parts **a** and **b** again. Draw a conclusion based on your results.

d) Any rectangle can be divided into a square and a smaller rectangle, as in the diagram.

Let the length of the shorter side of the original rectangle be 1 unit and the longer side be x units.

Label the rest of the sides of the diagram.

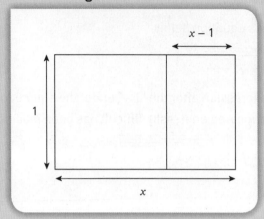

The object is to obtain a value for x such that the new smaller rectangle is similar to the original rectangle. Write an equation to satisfy this condition, and **solve** for x.

Comment on your result.

RESEARCH

The solution to the equations in **a, b** and **d** is represented by the Greek letter phi (Φ)

$$\Phi = \frac{1+\sqrt{5}}{2}$$

Phi is a special number called *the golden ratio*.
Write a research project on why this number is so special.

EXTENSION

Solve this endless equation.

$$x = 2 + \cfrac{1}{2 + \cfrac{1}{2 + \cfrac{1}{2 + \cfrac{1}{2 + \cdots}}}}$$

Comment on your result.
Can you find other numbers exhibiting the same result, in which a familiar number can be written as an endless fraction?

ATL SKILLS
Thinking
Consider multiple alternatives, including those that might be unlikely or impossible.

Fibonacci numbers

Fibonacci numbers 1, 1, 2, 3, 5, 8, 13, 21, … form a famous mathematical pattern characterized by a simple recurrence relation: $u_{n+1} = u_n + u_{n-1}$.

Fibonacci numbers also provide a simple way of approximating phi, the golden ratio.

Just take the quotient of two consecutive Fibonacci numbers.
$$\frac{1}{1} = 1; \ \frac{2}{1} = 2; \ \frac{3}{2} = 1.5; \ \frac{5}{3} = 1.66\ldots; \ \frac{8}{5} = 1.6; \ \frac{13}{8} = 1.625\ldots$$

Famous number patterns

Another number that can be written as an endless fraction is the natural number e. Write a short report on this number and **explain** why it is called the natural constant, and why it was given the symbol e.

Psychologists have confirmed that the golden rectangle is one of the most pleasing shapes to the eye. Artists often incorporate the golden rectangle into their paintings, sculptures or architecture. Look around your environment and try to find objects that are golden rectangles, or contain golden rectangles. You could take photographs of the object you find, measure the sides and see if the ratio of any two adjacent sides approximates phi.

In Leonardo da Vinci's *Mona Lisa*, a golden rectangle can be drawn around the face.

WEB LINKS
Search for the chimney of the Turku power station in Finland. It has the Fibonacci numbers on it in neon lights!

CHAPTER LINK
Chapter 7 on generalization has an activity on climbing stairs that is about Fibonacci numbers.

The Parthenon in Greece fits into a golden rectangle, and can be subdivided into golden rectangles.

TOPIC 2

Algebraic patterns

In this topic, you will use your knowledge of numbers and factorizing algebraic expressions to make conjectures about the patterns that develop. You will discover some useful properties to help you factorize trinomials in which the coefficients are large numbers.

👤 Activity 6   Investigating squares

a) Choose any two consecutive integers, then:

  (i) square each integer

  (ii) find the absolute value of the difference between the squares

  (iii) find the sum of the integers you started with.

What do you notice?

b) Choose any two integers that differ by two units, then:

  (i) square each integer

  (ii) find the absolute value of the difference between the squares

  (iii) find the sum of the integers you started with.

What do you notice?

c) Choose any two integers that differ by three units, then:

  (i) square each integer

  (ii) find the absolute value of the difference of the squares

  (iii) find the sum of the integers you started out with.

What do you notice?

d) Choose any two integers that differ by n units, then:

    (i) square each integer

    (ii) find the absolute value of the difference of the squares

    (iii) find the sum of the integers you started out with.

    What do you notice?

State a conjecture, using appropriate mathematical notation and terminology, based on the patterns you have noticed above.

Use algebra to justify your conjecture.

ATL SKILLS
Collaboration
Encourage others to contribute.

In the previous activity you found and described algebraic patterns in numbers. In this next activity, you will find them in the factorization of algebraic expressions

Activity 7   Investigating trinomials

In this investigation, you will consider trinomials of the form $ax^2+bx+c, a, b, c \in \mathbb{Z}^+, a \neq 0$

a) Factorize the following trinomials, showing all of your steps.

    (i) $6x^2+19x+10$

    (ii) $6x^2+23x+10$

    (iii) $6x^2+32x+10$

b) List all positive integer values of b (including the ones above), for which $6x^2+bx+10$ is factorizable. Then factorize each one.

c) How many of the trinomials above produced a highest common factor for 6, b and 10? Do not include 1.

d) You know that 60 has 12 factors. **Explain** how it follows that there are 6 values for b.

e) **Explain** whether interchanging a and c would change the possible values for b.

f) Form some new trinomials from those in part **a** by interchanging the numbers represented by a and c in the general form of the quadratic equation. **Explain** how the factors are related to the factors of the original trinomial.

g) Now, compare your results above by performing parts **b**–**f** on these trinomials.

    (i) $12x^2+27x+15$

    (ii) $12x^2+28x+15$

    (iii) $12x^2+29x+15$

Use the general form of the trinomial to summarize your findings. Make sure that you use appropriate mathematical notation and terminology.

ATL SKILLS
Collaboration
Encourage others to contribute.

TOPIC 3

Applying number patterns

In this topic you will consider some properties of real numbers to help you solve problems. If you can recognize patterns and make conjectures that work, you can save the time you would spend on experimentation. Your findings can be helpful in solving other problems.

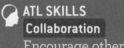
Activity 8   Mosaic tile patterns

Look at the design of these mosaic tiles.

If you look at the different stages in this design, you can work out the pattern that was used to create it.

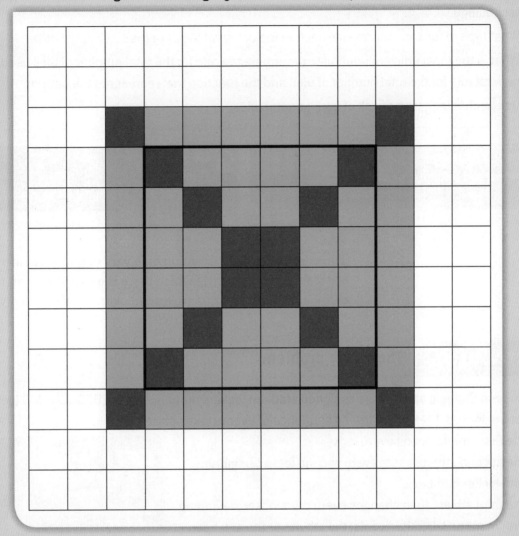

The first stage is enclosed in the black square, with each stage adding one more border of tiles around the outside.

Colour in stages three and four of the tiles.

Observe the design of the tiles. Look for patterns created by each colour.

Organize your information into the table below.

Stage number	Purple tiles	Green tiles	Orange tiles	Total number of tiles
1				
2				
3				
4				
n				

a) Determine the **general rule** for the number of tiles of each colour. Do not use a calculator or other technology.

b) What type of function represents each of the coloured tile patterns?

c) What is the most efficient method to determine the rule for the total number of tiles used? Find the general rule for the total number of tiles and the function that represents the pattern.

d) [Use] technology to verify all of your general rules.

🌐 **GLOBAL CONTEXTS**
Personal and cultural expression

🧠 **ATL SKILLS**
Thinking
Propose and evaluate a variety of solutions.

The next activity is a variation on a very common theme. You may also see it set as an activity in which there is a row of a given number of coins, and selected coins are turned over according to given rules.

 Activity 9 The locker problem

One thousand Olympic athletes are assigned stadium lockers numbered 1 to 1000. All 1000 lockers are shut, but not locked. Then the following sequence of events takes place.

- The first athlete opens every locker.
- The second athlete shuts every second locker, beginning with locker number 2.
- The third athlete then changes every third locker, beginning with number 3. In other words, the athlete shuts any open locker, and opens any shut locker.
- The fourth athlete only looks at every fourth locker, beginning with number 4. She shuts any open locker, and opens any shut locker.

🔗 **INTERNATIONAL MATHEMATICS**
When you write very large (or very small) numbers, with five or more digits either side of the decimal point, you set the digits out in groups of three, for example, 12 345.543 21. In some countries, commas or dots are used instead of the small spaces.

This continues until all one thousand athletes have followed this pattern with the 1000 lockers.

Investigate the patterns after the first 30 athletes have been at the lockers. Then answer these questions.

a) Which of the first 30 lockers are open and which are shut?

b) Which of the first 30 lockers were changed the most times?

c) Which of the first 30 lockers were changed exactly twice?

Now, try to generalize the patterns you saw and your results to answer these questions.

d) Which lockers will be open and which will be shut after the 1000 athletes have been at them? Will lockers 50, 100 and 1000 be open or shut?

e) If 10 000 athletes did the same exercise with 10 000 lockers, would the 10 000th locker be open or shut?

f) Which lockers have changed the most times after the first 100 athletes?

g) [Describe] the sequence of the numbers of the lockers that changed the most times by the 1000 athletes, without actually finding those locker numbers.

h) [Describe] the sequence of the numbers of the lockers that have changed exactly twice, without actually finding all those numbers.

🌐 **GLOBAL CONTEXTS**
Scientific and technical innovation

💭 **ATL SKILLS**
Thinking
Practise visible thinking strategies and techniques.

You will now move on to explore growth patterns, starting with one that occurs in nature.

 **Activity 10**   **Exponential growth and decay**

EXPONENTIAL GROWTH

When you study bacterial growth, you often look at how long it takes a generation of bacteria to double in number to make predictions about how fast the bacteria will spread and the effects.

> 🔗 **INTERDISCIPLINARY LINKS**   In biology, you learn that the majority of bacteria are harmless but some can make you sick (and can even be lethal). *E. coli* is a bacterium that lives in your intestine. Most strains have no adverse effect, but some can cause food poisoning and, in severe cases, death. A population of *E. coli* doubles in size every 20 minutes.

a) Calculate how many bacteria there will be in 10 generation cycles. Assume that the bacteria in this activity double in number every 20 minutes.

Fill in the table.

Number of generations (cycles)	0	1	2	3	4	5	6	7	8	9	10	x
Number of minutes	0											
Number of bacteria	1											
Pattern (connection between number of bacteria and number of generations)												

b) [Describe] in words the relationship between the number of generations and the number of bacteria.

c) Write a general rule (formula) to describe this relationship.

d) How many bacteria would there be after one day?

e) What would happen if you started with more than one bacterium?

Fill in the table, starting with 10 bacteria.

TIP

Put the data into your GDC and graph your results. Sketch the general shape, clearly showing any asymptotes.

Number of generations (cycles)	0	1	2	3	4	5	6	7	8	9	10	x
Number of minutes	0											
Number of bacteria	10											
Pattern												

f) [Explain] the key difference between the two patterns.

g) Write a general rule (formula) to describe this relationship.

EXPONENTIAL DECAY

One common form of exponential decay involves half-lives. The half-life of a substance is the amount of time it takes for it to decrease by half. For example, if a substance has a half-life of 5 hours and there are 20 g to begin with, then there will be 10 g left after 5 hours. You will know from your science lessons that matter does not just disappear. In this example, half of the substance has decayed to form a different substance.

DDT was an insecticide used all around the world. Its purpose was to control insects that carry diseases such as malaria and other insects that affected agricultural crops. It was used extensively by the Allies in the Second World War to limit typhus from spreading through the army camps. It has been banned in most countries since the 1970s because of its long-term harmful effects. It has a half-life of approximately 15 years. If a farmer used 40 pounds of active DDT, model its decay over time.

Number of cycles	0	1	2	3	4	5	6	7	8	9	10	x
Number of years	0											
Amount of DDT (lbs)	40											
Pattern												

a) [Describe] in words the relationship between the number of cycles and the amount of DDT.

b) Write a general rule (formula) to describe this relationship.

c) Put the data above into your GDC and graph your results. Sketch the general shape, clearly showing the asymptotes.

d) How much active DDT will there be after 300 years?

e) In general, a scenario involving exponential growth or decay can be modelled by $y=c(a)^x$.
Explain what each variable and parameter represents.

y is

x is

c is

The value of c always has to be _____

$a =$

f) Explain how the value of a will change, depending on whether you are dealing with exponential growth or decay.

g) Assuming all of these terms are positive, state the domain and range for these types of exponential function. Explain your reasoning.

REFLECTION

a) The news often reports *E. coli* outbreaks and food recalls. Given the nature of its growth, explain why particular *E. coli* strains can be so dangerous. Research some recent *E. coli* outbreaks. What can you do to avoid exposure to bacteria such as *E. coli* 0157:H7?

b) Why is the use of products like DDT so harmful to the environment?

c) Find three other examples in real life that can be modelled using exponential growth and three that can be modelled using exponential decay.

 **GLOBAL CONTEXTS**
Scientific and technical innovation

ATL SKILLS
Transfer
Make connections between subject groups and disciplines.

Now, from a consideration of something that is microscopically small, you will move on to consider the movement of celestial bodies in space.

 Activity 11 The Titius–Bode law

In 1766, Johann Titius made a conjecture about a relationship between the distances of the planets from the Sun. It wasn't until 1778 that Johann Bode expressed the relationship mathematically.

Take the distance between Earth and the Sun to be 1 astronomical unit. Then the distances of the other planets from the Sun, as proposed by the law, are recorded in column 3 of the table. Planets for which the distance from the Sun is less than one astronomical unit are between Earth and the Sun. They are closer to the Sun than Earth is.

Planets for which the distance from the Sun is greater than one astronomical unit are further from the Sun than Earth is. The planets' actual distances from the Sun are given in the right-hand column in each part of the table.

Planet	k	T–B rule distance (AU)	Real distance (AU)	Planet	k	T–B rule distance (AU)	Real distance (AU)
Mercury	0	0.4	0.39	Jupiter	16	5.2	5.20
Venus	1	0.7	0.72	Saturn	32	10.0	9.54
Earth	2	1.0	1.00	Uranus	64	19.6	19.2
Mars	4	1.6	1.52	Neptune	128	38.8	30.06
Ceres	8	2.8	2.77	Pluto	256	77.2	39.44

Where did this pattern come from? Is it a coincidence or is it planetary law? Work through this activity to express the sequence mathematically. Then try to determine the answer.

a) Titius–Bode's law begins with a simple sequence:

   0, 3, 6, 12, 24, 48, 96

   Write a recursive and an explicit formula for this sequence for $n > 1$ (starting with the second term).

b) If you add 4 to each of the above numbers, you get:

   4, 7, 10, 16, 28, 52, 100

   If possible, write a recursive and/or a general rule for this sequence for $n > 1$ (again starting with the second term). If not, try to write a formula to generate these terms. Use a new variable.

c) If you then divide each of the numbers in **b** by 10 you get:

   0.4, 0.7, 1, 1.6, 2.8, 5.2, 10

   If possible, write a recursive and an explicit formula for this sequence for $n > 1$ (again starting with the second term). If not, try to write a formula to generate these terms. Use a new variable.

d) Find the next three terms in the Titius–Bode law and compare them to the values in the table.

e) In 2006, Pluto was demoted to being a dwarf planet. Do you think this was the correct decision? Explain.

f) The planet Ceres is really just a large asteroid, so many scientists refuted the law based on this. Recently, however, the Titius–Bode law has been found to hold for planets orbiting other stars. Research this discovery and explain whether you think it is just a coincidence or a natural pattern in planetary systems.

GLOBAL CONTEXTS
Scientific and technical innovation

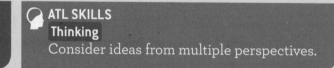
ATL SKILLS
Thinking
Consider ideas from multiple perspectives.

In this last activity you will attempt to find order within a seemingly random pattern of numbers.

 Activity 12 The chaos game

Work with a partner to do this activity. You will need a normal six-sided die and transparent paper.

Draw a large equilateral triangle and label the vertices L (left), R (right) and T (top).

A random roll of the die will determine the direction for each move. Choose any arbitrary point inside the triangle as your starting point. Then, successive rolls of the die will determine the direction that you will move from this starting point, halfway to the designated vertex.

Assign L to faces 1 and 2, T to faces 3 and 4, and R to faces 5 and 6 of the die.

Your first move is to mark any point inside the triangle with a 1, and roll the die.

Plot the point halfway between your starting point and the vertex named by the die. Mark this new point with 2.

Roll again and plot the next point halfway between the new point and the vertex indicated by the roll.

As you play the game, mark each new midpoint with the number of the move.

Make at least 25 moves.

Can you identify an emerging pattern? **Describe** it.

Collect all the class transparencies and place them one on top of the other. Can you see a pattern now?

QUESTIONS

a) What is the boundary for all possible points in this game?
b) Do the successive midpoints appear to be randomly located within the boundary, or not? **Explain** .
c) What appeared to be random actually has an invisible pattern. **Explain** .

ATL SKILLS
Thinking
Consider ideas from multiple perspectives.

Reflection

a) In this topic, you have seen a variety of number patterns in different real-world contexts. What other number patterns exist in your daily life? How can they be described mathematically?

b) You have also seen number patterns in mosaics and in planetary systems. Do you think these patterns occur by accident? **Explain** .

c) Research the Sierpinski triangle. **Explain** the connection between the chaos game and this famous triangle.

Summary

In this chapter, you have looked at some famous patterns that have intrigued mathematicians throughout the ages. You have also seen where many of these patterns can be found in nature, and indeed in your everyday life. You have investigated patterns algebraically and within the context of real-life problems.

When solving a complicated problem, it might help to see if there are hidden patterns that could suggest a strategy to solve it. Patterns are everywhere, and recognizing them is usually the first step to solving a problem, whether it is an exercise you have to do for school or related to your future career.

CHAPTER 12 Quantity

An amount or number

INQUIRY QUESTIONS	**TOPIC 1** Volumes, areas and perimeters
	■ Is there such a thing as the perfect solution?
	TOPIC 2 Trigonometry
	■ How can the choice of unit for a quantity affect how you study it?
	TOPIC 3 Age problems
	■ Can classification of quantities be an ambiguous process?

SKILLS

ATL

✓ Propose and evaluate a variety of solutions.

✓ Create novel solutions to authentic problems.

✓ Make inferences and draw conclusions.

✓ Select and use technology effectively and productively.

✓ Make unexpected or unusual connections between objects and/or ideas.

✓ Develop contrary or opposing arguments.

✓ Understand the impact of media representations and modes of presentation.

Algebra

✓ Determine and manipulate expressions of one variable in terms of other specific variables given, with and without a GDC.

Geometry and trigonometry

✓ Calculate the dimensions of shapes, given some conditions, and determine values that maximize them.

✓ Use degrees and radians to express sizes of angles and recognize the advantage of the use of each unit.

✓ Graph sinusoidal functions and analyse the effects of parameters of these functions on their graphs.

✓ Reflect on the meaning of parameters in a real-world context.

Statistics and probability

✓ Calculate mean, mode and median of student ages.

✓ Conduct a graphical analysis of histograms.

✓ Recognize differences between continuous and discrete ungrouped and grouped data; in particular, analyse the impact of treating age as a discrete or continuous variable.

OTHER RELATED CONCEPTS

Measurement   Representation   Space

GLOSSARY

Sinusoidal function named after the sine function, of which it is the graph, a specific type of periodic function that describes a smooth, repetitive oscillation and produces graphs that look like waves, where any portion of the curve can be translated onto another portion of the curve.

COMMAND TERMS

Describe give a detailed account or picture of a situation, event, pattern or process.

Discuss offer a considered and balanced review that includes a range of arguments, factors or hypotheses. Opinions or conclusions should be presented clearly and supported by appropriate evidence.

Justify give valid reasons or evidence to support an answer or conclusion.

Use apply knowledge or rules to put theory into practice.

Introducing quantity

In mathematics the concept of quantity is an ancient one, extending back to the time of Aristotle and earlier.

🏛 MATHS THROUGH HISTORY

Aristotle

Aristotle is one of the most learned scholars of all times. This ancient Greek philosopher was born circa 384 BC in Stagira, Greece. When he turned 17, he enrolled in Plato's Academy. In 338 BC, he began tutoring Alexander the Great. In 335 BC, Aristotle founded his own school, the Lyceum, in Athens, where he spent most of the rest of his life studying, teaching and writing. Aristotle was primarily a philosopher and biologist but he also applied himself to mathematics. In particular he studied the controversial Zeno paradoxes and the concept of infinity. Aristotle is also considered the father of mathematical logic. Aristotle died in 322 BC.

Quantity can exist as a magnitude or as a number. It is among the basic types of property studied in mathematics alongside quality, change and relation. As a fundamental term, quantity is used to refer to any type of numeric property or attribute of something.

In simple terms, quantity is defined as an amount that is measurable in dimensions, units or in extensions of these, which can be expressed in numbers or symbols. It is anything that can be multiplied, divided or measured. For example, time is a type of quantity. It can be measured in hours, minutes and seconds. However, colour is not a quantity. It cannot be said that one colour is twice as great, or half as great, as another. So, a quantity is something that has—or is measured in—units, such as three pencils or four and one-half metres. A number is an abstract concept that represents a particular quantity. Some examples of quantities are:

- height
- depth
- area
- length
- width
- volume
- time
- speed.

It is important to note that a quantity can be variable or constant. It can also be known or unknown. In mathematics you use specific words such as "variable", "parameter" and "constant" to describe mathematical quantities. When solving problems you may need to define appropriate quantities or choose the scale and the origin in graphs and data displays. If you want to model a real-life situation you also need to choose an appropriate level of accuracy and consider limitations on measurements when reporting quantities.

In this chapter, you will explore different types of quantities used in various areas of mathematics to help you to understand this concept better.

TIP

Find out more about parameter in Chapter 13, Representation.

Volumes, areas and perimeters

Considering the connection between quantity and geometry is a good starting point, if you want to understand the concept of quantity. Comparing numerical quantities helps you to recognize the relationship between them. In geometry, you can start with formulas that relate the properties or dimensions of 2D and 3D shapes. Next, you can choose one of the quantities and change its magnitude, to see how this affects all of the other quantities.

It is essential that you understand the different units of measurement, and the relationships between them, for each of the numerical quantities. You must also be able to convert between units, when working on geometry problems.

Activity 1   **Maximizing area**

Start by looking at a 2D problem. Find the maximum area that can be enclosed by a rectangle with a fixed perimeter of 60 cm.

a) To help visualize this task, sketch a couple of rectangles with a perimeter of 60 cm and calculate their areas.

b) Now create a table so that you can look at all possible lengths and widths to determine which will give the maximum area.

c) What degree of accuracy do you think is needed for the quantities?

d) What can you say about the relationship between the values of the dimensions of the rectangle and the area?

e) What do you notice about the rectangle that has the maximum area?

f) Would these circumstances hold true for any given perimeter length? **Justify** your answer by testing other perimeter lengths.

g) In what sort of real-life applications would this sort of information be helpful?

ATL SKILLS
Thinking
Propose and evaluate a variety of solutions.

You can now use a similar problem-solving process to help you solve more complex problems involving numerical quantities and geometry.

Consider this scenario:

A large paper-manufacturing company has asked you to create a template for paper cones. These are to be used by people using water coolers in offices, gyms and so on. You have been given a circular press with a radius of 9 cm. Each paper cup must hold as much water as possible.

STEP 1   **Explore different templates.**

You have been assigned a piece of cardboard. Cut out a circle of radius 9 cm, then remove a circular sector. Note the angle of the sector. Bend the remaining circular sector to form a hollow cone. The straight sides of the sector should meet without overlapping, as in the diagram below.

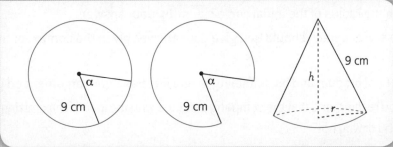

a) Find the height of your cone and the radius of its base and then calculate its volume.

b) Tabulate everyone's results in a table like the one below.

Sector angle, $\alpha°$	Radius of the base, r cm	Height of the cone, h cm	Volume of the cone, V cm³

c) What can you say about the relationship between the value of α and the volume of the cone?

d) Based on everyone's results, what angle do you think will produce the greatest volume?

STEP 2   **Represent the volume algebraically and graphically.**

Next, working on your own, write a report to the company. Follow the instructions below.

a) Find an expression for r in terms of α.

Remember—when using the sector formula you are considering the sector that is cut out and your units are degrees.

b) Find an expression for h in terms of α.

There is no need to simplify your expression fully, as you will be using technology to arrive at the result.

c) Find an expression to represent the volume of the cone.

d) Use a GDC or graphing software to:
- graph the volume of the cone in terms of α (Take a screenshot of this to put in your report.)
- calculate the value of α that maximizes this volume. (Take a screenshot of this to put in your report.)

e) [Discuss], in detail, the characteristics and key points of your graph and the changes in volume that result from changes in the size of the sector angle.

f) Draw a labelled prototype of the conical paper-cup template.

a) In mathematics, sometimes you only need to look at part of a function when graphing it, given the context of the question. For this activity, are there any restrictions on the domain that you would have to consider, so that it makes sense in a real-life context? If so, [discuss] what those restrictions would be.

b) Do you think the size of the sector angle that produces the maximum volume would change, depending on the radius of the initial circle? [Justify] your answer.

c) To what degree of accuracy should you give your answer, given the context of the problem? [Justify] your answer.

d) [Discuss] the final product and its suitability in real life. Base your argument on your calculations.

  i) Would you recommend that the company continue with the idea of maximizing the volume of each cone? Why or why not?

  ii) Would you change the template in any way for the manufacturing stage? If so, what would you do?

 **GLOBAL CONTEXTS**
Scientific and technical innovation

 **ATL SKILLS**
Thinking
Create novel solutions to authentic problems.

TOPIC 2

Trigonometry
Why use radians to measure angles?

Have you ever noticed that your GDC's default unit for angle measure is radians?

If you investigate further you will discover that the radian is the SI unit for plane angles.

But why did the scientific community adopt the radian instead of the degree as the preferred unit when it comes to measuring angles? After all, everybody seems to use degrees to measure angles in geometry!

Look up the relation between radians and degrees. You will discover that:

$$1 \text{ rad} = \frac{180°}{\pi} \approx 57.2958°$$

In previous years, you have used a protractor to measure angles. This allowed you to obtain very accurate measurements because 1° is represented by a very small arc on the edge of the protractor. So why do scientists use a unit that corresponds to a much bigger arc? Is there any advantage in choosing a unit that may lead to less accurate results?

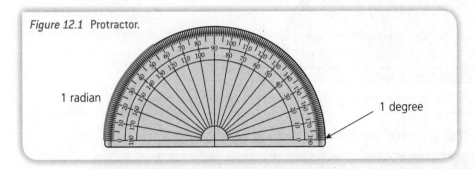

Figure 12.1 Protractor.

1 radian

1 degree

If you ask a mathematician, they might say: "It is quite simple. The radian is a dimensionless quantity. This means that the radian is a pure number that needs no unit symbol, and mathematicians would almost always omit the symbol 'rad'. In the absence of any symbol, the angle is assumed to be in radians. When they want to express an angle in degrees, they use the symbol ° to write it."

🏛 MATHS THROUGH HISTORY

The term "radian" first appeared in print in 1873, in examination questions set by James Thomson at Queens College, Belfast. However, the concept of radian measure is normally credited to Roger Cotes in 1714, who defined the radian in everything but name. About 300 years before that, the Persian mathematician and astronomer Al-Kashi used so-called "diameter parts" as units where one diameter part was $\frac{1}{60}$ radian. Al-Kashi is credited with having produced very accurate sine tables, used to solve triangle problems long before calculators were invented.

🔗 INTERNATIONAL MATHEMATICS
In general, when the size of an angle is given exactly as a multiple of pi (for example, $\frac{3}{4}\pi$), no symbol for units is used. However, in some countries if the value is expressed in decimal form, a superscript c (that stands for circular measure) is often used to indicate that the unit in use is the radian.

The mathematicians' explanation may not convince you, but the fact that the radian is the SI unit gives you a good reason to explore the topic further.

In the next activity, you will discover some advantages of using the radian. You will learn how you can define the trigonometric ratio, sine, as a function just by looking at the movement of a point on a bike tyre!

You will need several protractors of different sizes.

STEP 1   Draw an angle on a sheet of white paper. Use different protractors to measure the angle a (in degrees) and, in each case, measure also the diameter, d, of the protractor and the length l of the corresponding arc.

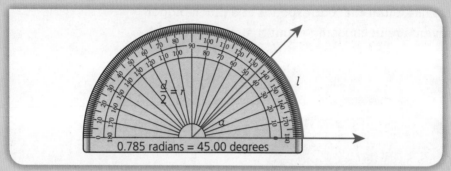

0.785 radians = 45.00 degrees

STEP 2   Repeat for other angles of your choice and complete the first five columns of this table.

a (degrees)	l	d	$r = \dfrac{d}{2}$	$\dfrac{l}{r}$	a (radians)

STEP 3   **Use** the conversion factor 1 rad corresponds to $\left(\dfrac{180}{\pi}\right)^{\circ}$ to calculate the size of the angle in radians and complete the table.

STEP 4   Explain how your results support the statement "the radian is dimensionless". Write down a definition of radian, using the terms "ratio", "length of the arc, l" and "radius, r".

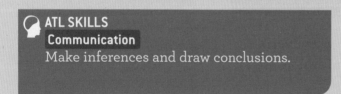

ATL SKILLS

Communication

Make inferences and draw conclusions.

In the previous activity, you may have noticed an interesting relation between a (in radians) and l for the case when $r = 1$ unit and $l = a$. An angle of one radian at the centre of the circle cuts off an arc equal in length to the radius of the circle.

This means that if you use the radius of the protractor as the unit of length, the angle can be measured in terms of a length too. This can be very useful when you are solving practical problems but you do not have available protractors of convenient sizes. For example, suppose that you want to study the movement of a bike tyre. The tyre can be modelled by a circle if you ignore its thickness. If you mark a point on the tyre and measure the time taken for the point to complete a certain number of revolutions, you can calculate the speed of the bike. This is the ratio between the number of revolutions and the time taken to complete them. For example, one revolution per second is equal to 2π radians per second.

Figure 12.2. Studying the movement of a bike tyre.

1 unit

1 revolution = 2π units

◯◯ INTERDISCIPLINARY LINKS Physicists often use angular quantities when describing sound waves or the motion of objects in circles. For example, when describing the journey taken by a point on a bicycle wheel, using radians instead of degrees makes it much easier to work out other quantities such as the velocity or the acceleration of the object. So, in physics, the SI unit "radians per second" is widely used for angular measurement.

In many branches of mathematics that you will study during your school career, apart from practical geometry, you, too, will measure angles in radians. This is because radians have a mathematical "naturalness" that leads to a more elegant formulation of a number of important results involving trigonometric functions.

Motion that repeats itself over equal time intervals is called *periodic motion*. The Earth's orbit, tides and even the vibration of guitar strings are all periodic because the movement occurs over and over again during the same interval of time. When quantities change in this way, their graphs also demonstrate this cyclic behaviour.

You are going to investigate a sinusoidal function, with the help of an everyday object. You will need a bicycle tyre. It can still be on the bike.

a) Position the tyre so that the valve on the inner tube (where the air goes in) is at the top. Measure the height of the valve from the ground. Rotate the tyre and measure the height of the valve above the ground at regular intervals. Do this until you have made one complete revolution. Take at least 10 measurements. Make a table of your results.

b) Graph your data. Use an appropriate scale.

c) What would your graph look like, if you continued taking measurements for more revolutions? Draw two more of these cycles on your graph.

d) Find out what the words "amplitude", "period" and "line of equilibrium" mean, in relation to sinusoidal functions. Indicate them on your graph.

Now consider two variable quantities: x is the independent variable that represents the size of an angle, in radians; and y is the dependent variable defined by the relation $y = \sin x$.

e) Draw up a table of values, like the one below. Use your calculator to find the values of y. Draw a graph of the function $y = \sin x$.

x	$y = \sin x$	x	$y = \sin x$
0		$\dfrac{5\pi}{4}$	
$\dfrac{\pi}{4}$		$\dfrac{3\pi}{2}$	
$\dfrac{\pi}{2}$		$\dfrac{7\pi}{4}$	
$\dfrac{3\pi}{4}$		2π	
π			

f) Determine the amplitude, period and line of equilibrium for this sine function.

g) Explain how the two graphs you have drawn in this activity are related to one another.

 **GLOBAL CONTEXTS**
Scientific and technical innovation

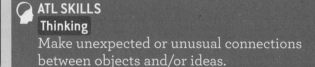 **ATL SKILLS**
Thinking
Make unexpected or unusual connections between objects and/or ideas.

The basic sine curve occurs in many forms and in many applications. It is often modified by the application of parameters. In the next activity you will investigate how its shape changes, when these parameters change.

In this activity, you will experiment with various parameters of a family of sinusoidal functions, study their effects on their graphs and how these can be applied in design. You will be using a graphic display calculator, either handheld or online, and changing each of the parameters (a, b and c) in the function $y = a \sin bx + c$.

By changing these parameters one at a time you can see their individual impact on the overall graph before applying several of them together.

STEP 1 Study the effect of the value of the parameter a on the graph. First make a conjecture (write it down) and then test it by filling in a table similar to the one below. Copy the graphs from the calculator and write your conclusion based on your observations. Make sure that you use appropriate mathematical terminology in your writing.

Conjecture: "Changing the value of a will affect _____."

Function	Graph
$y_1 = \sin x$	
$y_2 = 2 \sin x$	
$y_2 = 5 \sin x$	
$y_2 = \frac{1}{2} \sin x$	
$y_2 = -3 \sin x$	

STEP 2 **Describe** how changing the value of the parameter a affects the graph of $y = a \sin x$.

STEP 3 Conduct similar experiments to explore the effects of changing parameters b and c on the graph of $y = \sin bx$ and $y = \sin x + c$. Remember to make a conjecture first, test it with an experiment and then write the conclusion. Be sure you are only changing **one** quantity each time.

STEP 4 Test your conjectures.

Describe the effects of each parameter on the graph of $y = \sin x$ and then sketch the graph.

i) $y = -3\sin x + 2$ **ii)** $y = \sin 4x - 5$ **iii)** $y = 2\sin\dfrac{1}{4}x + 3$

REFLECTION

a) Suppose you measured angles in degrees. How would that affect your graphs?

b) None of the transformations seen in this section produce a horizontal translation. What parameter would you need to introduce in order to translate the sine graph to the left or right?

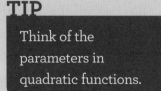

TIP

Think of the parameters in quadratic functions.

 ATL SKILLS

Self-management

Select and use technology effectively and productively.

Before you start the next activity, look back at Activity 4 to remind yourself of what you did.

 Activity 6   **Changing a tyre**

Use your results from Activity 4 to answer these questions.

a) Find an equation for the graph.

b) What change would you have to make to the tyre, to change the amplitude of your graph? How would that affect the bicycle and/or its performance?

c) What would you have to change on the tyre, to change the line of equilibrium of your graph? How would that affect the bicycle and/or its performance?

d) What would you have to do to change the period of your graph? How would that affect the bicycle and/or its performance?

REFLECTION

a) What other phenomena are periodic? Can they be described by sinusoidal functions, or are there other periodic functions?

b) Research Galileo and the cycloid. How does the graph of the cycloid relate to the movement of a point on a bike tyre? What other interesting properties does a cycloid have?

 GLOBAL CONTEXTS

Scientific and technical innovation

 ATL SKILLS

Thinking

Make unexpected or unusual connections between objects and/or ideas.

Age problems

Quantities can be either discrete or continuous. Measures such as angle size, length, area and volume are examples of continuous quantities. They can take any value in an interval. In theory, they can be measured to any degree of accuracy.

Numbers of people or objects are discrete quantities, you can just count them and use natural numbers to represent them. You often use discrete quantities such as numbers of people when you solve problems. But what about quantities such as age? Is age a discrete quantity or a continuous quantity? What are the implications of this decision when you are graphing and determining statistics about age data?

 Activity 7   **Mean, mode and median age**

Complete a table like this one about the ages of the students in your class.

Birth at (year, month, day, hour, minute)	Age in number of years completed	Age in number of days lived so far	Age in number of minutes lived
(for example, 1999 Jan 19, 23:10)			

a) Use the age data you have collected to calculate the mean, the mode and the median age of the students in your class.

b) Compare the results for each average: mean, mode and median.

c) Comment on the results. **Discuss** the advantages and disadvantages of working with age rounded to the "completed number of years". Refer to a real-life context: how often is age treated as a discrete quantity?

d) List other quantities that can be treated both as continuous or discrete. In each case, give reasons in favour of and against each choice.

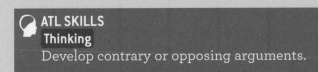
ATL SKILLS
Thinking
Develop contrary or opposing arguments.

Discrete and continuous quantities

In statistics, you use different types of bar graph to represent discrete and continuous quantities.

Usually discrete quantities are represented by bar charts like these.

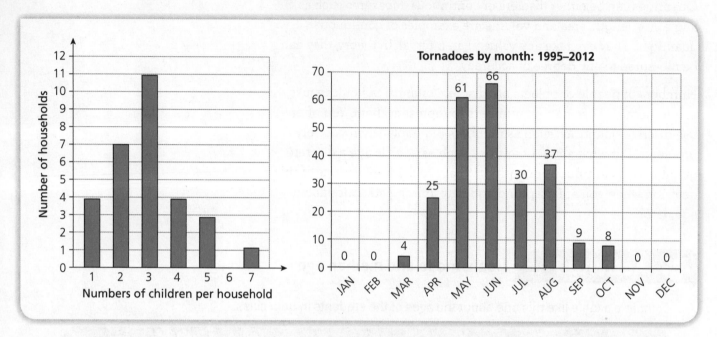

Bar charts are very simple graphs. They are easy to read as you just need to look at the height of each bar to read the frequency of each value of your discrete variable. Both the width of the bars and the space between them are irrelevant. Some bar charts have no spaces between the bars. Sometimes the bars have no width, they are just vertical lines.

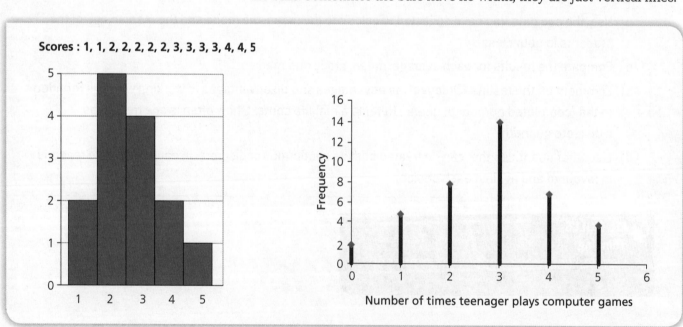

When you want to represent continuous quantities you should use histograms. This type of graph can be more complex to use or read as the frequency of each class or interval needs to be proportional to the area of the corresponding bar.

Create your own visual dictionary of key vocabulary in statistics. Write the appropriate definition, then illustrate each word with pictures of real-life examples (taken by you or from newspapers, magazines, and so on). You can then refer to this whenever you need to.

Activity 8   Histograms

The histogram below shows information about the ages of people at a small beach. The age groups are $a < 15$, $15 \leqslant a < 25$, $25 \leqslant a < 40$ and so on.

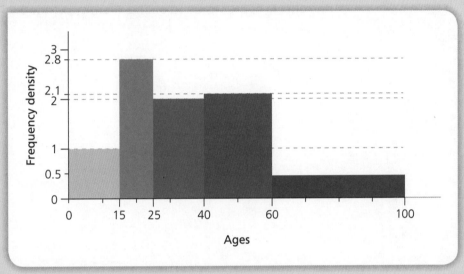

a) How many people were under 15 years old?

b) Which age group (or class) is the modal class? This is the age group with the highest frequency.

c) How many people were at the beach?

d) Comment on the difficulty of using the given histogram to answer the questions above.

e) Comment on the accuracy of the information provided by the histogram with reference to the real-life situation. For example, if someone answers a survey question such as: "How old are you?" with a simple 15, how accurate is this answer?

REFLECTION

Summarize the differences between discrete and continuous data. What do you think is the most appropriate way to display each? Explain.

 GLOBAL CONTEXTS
Personal and cultural expression

 ATL SKILLS
Research
Understand the impact of media representations and modes of presentation.

Summary

At the most basic level, mathematics is often seen as the science of numbers. The Pythagoreans even believed that "all is number". By this, they meant that everything can be quantified, either by counting discrete quantities, or by measuring continuous quantities.

In this chapter, you have explored both types of quantity. You then went further and thought about refined classifications of quantities: parameters versus variables, or quantities that are related by relationships and are classified as independent and dependent. You also considered ways to represent quantities graphically and explored the effects of the variation of one quantity on another.

13 Representation

The manner in which something is presented

INQUIRY QUESTIONS

TOPIC 1 Points, lines and parabolas
- ■ **How does representation influence interpretation?**

TOPIC 2 Probability trees
- ■ **Is representation a form of simplification?**

TOPIC 3 Misrepresentation
- ■ **Is any representation unbiased?**

SKILLS

ATL
- ✓ Consider ideas from multiple perspectives.
- ✓ Revise understanding based on new information and evidence.
- ✓ Evaluate and manage risk.
- ✓ Understand the impact of media representation and modes of presentation.

Number
- ✓ Use points on the parabola and the y-axis to generate the times table.

Algebra
- ✓ Graph parabolas.
- ✓ Use set-builder notation.
- ✓ Represent points and lines graphically.
- ✓ Solve problems involving position, velocity and time and intersection of paths of moving objects when represented by Cartesian and vector equations of lines.

Statistics and probability
- ✓ Use probability trees to represent a situation and calculate the probability of selected outcomes.
- ✓ Use probability trees to calculate the conditional probability of a real-life situation.
- ✓ Use and interpret measures of central tendency and measures of dispersion.
- ✓ Interpret graphs.

OTHER RELATED CONCEPTS

Quantity   Simplification

GLOSSARY

Parameter a quantity of which the value is selected for the particular circumstances and in relation to which other variable quantities may be expressed.

Velocity the speed of something in a given direction.

COMMAND TERMS

Describe give a detailed account or picture of a situation, event, pattern or process.

Explain give a detailed account including reasons or causes.

Use apply knowledge or rules to put theory into practice.

Introducing representation

Mathematical representation is a term that refers to all of the different ways in which you can show a mathematical concept or relationship. You can represent your mathematical ideas in a clear and visual way. You can use different symbols, a number or algebraic equation, a diagram, chart or a graph. Representation also includes the way that you visualize a mathematical idea as you work through it in your head. Representations help you to develop and record your mathematical thinking.

The choice of how to represent mathematical ideas can significantly affect the problem-solving process and the ease with which it can be communicated to others. Good notation, well-defined terms, and clear, well-labelled diagrams can make the difference between a readable mathematical piece of work and one that is confusing and difficult to follow. Even something as simple as how you write a rational number (fractional or decimal form) will have an impact on how easily you can work with it to analyse and solve a problem and communicate this solution.

Developments in technology and its availability in the classroom have brought a wider range of possibilities to represent data. You can now use spreadsheets and dynamic software to plot graphs and manipulate them easily. You can produce simulations and create a variety of scenarios in ways that are impossible by hand. You can also be creative, combine tools and even work in collaboration with your peers to produce amazing presentations of your findings when exploring complex problems. More than ever your classroom activities prepare you for life. It is not enough to find an answer: in a complex situation you will have to suggest possible scenarios and solutions and justify your solution. This is true whether you are an engineer, an architect, an economist or work in any other area.

In this chapter, you will explore links between different areas of mathematics and various ways of representing the same mathematical objects. You will also reflect on the advantages that each form of representation has when solving particular problems.

> *Mathematics is a linguistic activity; its ultimate area is preciseness of communication.*
>
> William L. Schaaf

INTERNATIONAL MATHEMATICS

Mathematics is said to be the universal language. Many cultures use a base 10 number system but represent the numbers in very different ways. Do some research on different forms of mathematical symbols and forms that are used in various cultures. Determine and perhaps debate if mathematics is indeed the universal language.

WEB LINKS

Go to http://ed.ted.com and search for "a clever way to estimate enormous numbers". How did Fermi represent numbers so that he could make rapid estimations with seemingly little data? Based on this idea, how could you accurately estimate the number of sweets in a jar (and perhaps win that jar as a result!)?

TOPIC 1

Points, lines and parabolas

You are familiar with the classic representation of real numbers on the real number line. In the first activity, you will discover how to use a parabola to represent the real numbers graphically and, in particular, the prime numbers, by using points on the parabola and the *y*-axis to generate the times table.

This is the graph of $y = x^2$. Two points on the graph are labelled, and a line segment connects the two points.

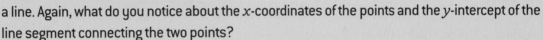

STEP 1 **a)** What do you notice about the relationship of the x-coordinates of the points and the y-intercept of the line segment connecting the two points?

b) Label any two other points on the graph on either side of the y-axis, where the x-coordinate is an integer. Connect the two points with a line. Again, what do you notice about the x-coordinates of the points and the y-intercept of the line segment connecting the two points?

c) Repeat the above, selecting five different pairs of points, on opposite sides of the y-axis. What do you notice?

d) Does it matter if you choose integer values, or does your observation also hold when x is not an integer?

e) Can you explain why the graph above is called a parabolic times table? Why does it work?

STEP 2 Draw the graph of $y = x^2$ using a window of $[-5, 5]$ for the x-axis, and $[-1, 26]$ for the y-axis. Consider all the points in which the x-coordinate is greater than 1 or less than -1. Draw line segments between pairs of points. (Use integer values only.) What numbers are left untouched on the y-axis? What is special about these numbers?

STEP 3 Draw a graph of $y = x^2$, and extend the y-axis to -26. Choose two points whose x-coordinates are either both positive integers or both negative integers, and again draw a line segment between the two points. What is the relationship between the x-coordinates and the y-intercept of the line?

REFLECTION

Do some research on the Sieve of Eratosthenes, then practise using the algorithm to find the prime numbers less than 50. **Explain** why this algorithm is limited in its usefulness for finding prime numbers.

WEB LINKS
For an interactive sieve of Eratosthenes, go to http://www.hbmeyer.de and follow the link to Eratosthenes' prime sieve.

GLOBAL CONTEXT
Scientific and technical innovation

ATL SKILLS
Thinking
Consider ideas from multiple perspectives.

Points and straight lines are the simplest geometrical objects that you know. Geometrically a point is simply represented by a dot. Algebraically, when working in 2D, you can represent a point by an ordered pair of numbers once you have chosen a pair of Cartesian axes. In algebra, these are labelled the x-axis and the y-axis. Straight lines can be represented by segments drawn along a straight edge. You can think of these lines as extending in both directions, endlessly. Algebraically, a line can be seen as a continuous set of points with coordinates (x, y) that follow a certain rule given by a linear equation.

The next activity leads you to look closely at equations of lines and further explore their meaning.

👤 Activity 2 — Graphical representation of lines

STEP 1   For each point (x, y) on the grid, find the value of $x + y$.

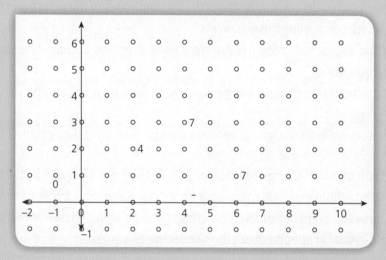

Use different colours to represent each of the following sets of points on the grid.

$A = \{(x, y) \mid x + y = 4\}$

$B = \{(x, y) \mid x + y = 6\}$

$C = \{(x, y) \mid x + y = 7\}$

$D = \{(x, y) \mid x + y = 1\}$

Describe the graphs that you obtain. What is the minimum number of points you need to plot, to obtain a graphical representation of each of the sets above?

STEP 2   **Explain** the meaning of the statement: "Two points define a line." **Explain** clearly how to draw the graph of a linear function efficiently.

     a)   **Use** your method to draw the graph of $2x + y = 6$. Solve the equation for y and use your calculator to confirm your answer.

     b)   On the same axes, graph the line with equation $y - x = 0$. Find the coordinates of the point where the two graphs intersect. **Use** your GDC to verify your answer.

STEP 3 Plot the points A(1, 2) and B(3, 5) and find an equation of the straight line AB:

    a) by determining the values of the parameters a and b in the equation $ax + by = 1$

    b) by determining the gradient and y-intercept of the line

    c) using regression and your calculator.

STEP 4 Apply your knowledge to solve this problem.

The table shows some equivalent temperatures in °C and °F.

Temperature x (°C)	20	37	55	70	100
Temperature y (°F)	68	98.6	131	158	212

    a) Plot the points to represent these pairs of temperatures.

    b) Find the relation between the variables x and y. Write it as a formula in which y (°F) is the subject.

    c) Transpose the equation so x (°C) is the subject.

    d) Use these equations to:

        i) determine the temperature in °F that corresponds to 0°C

        ii) determine the temperature in °C that corresponds to 23°F

        iii) find the temperature that is numerically the same in both scales.

STEP 5 Consider the points A(1, 2), B(3, 6) and C(4, 11).

    a) Is there any straight line that contains all three points? Explain.

    b) Find a line or a curve that contains all three points. Explain your strategy.

GLOBAL CONTEXT
Scientific and technical innovation

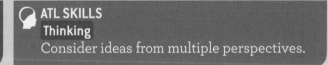

ATL SKILLS
Thinking
Consider ideas from multiple perspectives.

Cartesian equations of lines

Cartesian equations can be represented by lines on a set of Cartesian axes. You can use them to solve problems involving two variables. You can determine the value of one variable when you know the value of the other one. Functions that you are familiar with usually contain two variables, the independent and dependent variables.

Some unknowns do not necessarily vary. These are called **parameters.** For example, in the equation of the straight line, $y = mx + c$, x is the independent variable and y is the dependent variable. For any given straight line, you can determine the gradient (m) and y-intercept (c) and they don't vary within a straight line. These are the parameters of the function.

Parameters are extremely powerful tools for mathematicians and scientists in general, as they act neither as constants nor variables.

You can change the value of parameters repeatedly, to examine the relation between the variables. Changing the parameters allows for detailed analysis of the relation between variables. You can use parameters to simulate real-life situations that change frequently. The most common parameter you encounter in school mathematics is time.

Consider that you are studying the movement of an object along a straight line. For simplicity, let the object be represented by a point, called P, on the line. After selecting a suitable pair of axes, you could represent the initial situation in a diagram like this one.

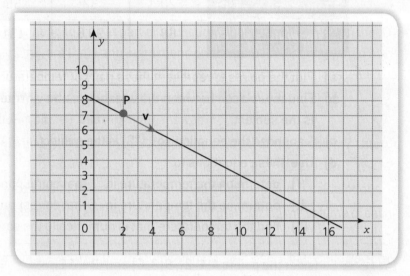

In this case, the point P has initial coordinates $(2, 7)$ and its initial position vector is $\overrightarrow{OP} = \begin{pmatrix} 2 \\ 7 \end{pmatrix}$. Now, suppose that this point is moving along the line in the direction shown by the **velocity** vector **v**. Then the point moves two units to the right and one unit down for each unit of time (for example, each second)—this means that $\mathbf{v} = \begin{pmatrix} 2 \\ -1 \end{pmatrix}$. If you want to find the positions of the point after one, two or three seconds you need to consider the initial position, the velocity vector and time.

So, after one second, P moves to position P_1 given by $\overrightarrow{OP_1} = \begin{pmatrix} 2 \\ 7 \end{pmatrix} + \begin{pmatrix} 2 \\ -1 \end{pmatrix} = \begin{pmatrix} 4 \\ 6 \end{pmatrix}$;

after two seconds, P moves to position P_2 given by $\overrightarrow{OP_2} = \begin{pmatrix} 2 \\ 7 \end{pmatrix} + 2\begin{pmatrix} 2 \\ -1 \end{pmatrix} = \begin{pmatrix} 6 \\ 5 \end{pmatrix}$

and after three seconds, P moves to position P_3 given by

$\overrightarrow{OP_3} = \begin{pmatrix} 2 \\ 7 \end{pmatrix} + 3\begin{pmatrix} 2 \\ -1 \end{pmatrix} = \begin{pmatrix} 8 \\ 4 \end{pmatrix}$. Both the graph and the vector expressions show a clear relation between the pair of coordinates of the position of the point, the constant vector that represents the velocity and the parameter time, which you can label t. This relation can be described by a vector equation: $\overrightarrow{OP} = \begin{pmatrix} 2 \\ 7 \end{pmatrix} + t\begin{pmatrix} 2 \\ -1 \end{pmatrix} = \begin{pmatrix} x \\ y \end{pmatrix}$. This is an equation of the line on the diagram but provides different information about the line than a Cartesian equation of the same line would. In this case, the vector equation describes the position of a moving point on this line when you know an initial position, its constant velocity and you choose the value of the parameter t. Of course, the information available allows you to find a Cartesian equation of the line $y = -\dfrac{1}{2}x + 8$, as the velocity vector determines the slope (or gradient) and the y-intercept is shown on the diagram.

The next activity leads you to explore the differences between Cartesian and vector lines in context and you are challenged to create a simulation of the situation given!

INTERDISCIPLINARY LINKS Many quantities in physics, such as force, velocity and acceleration, are vector quantities. They differ from scalar quantities in that they have both magnitude (or size) and direction. Physics students need to be able to perform operations with vectors, to analyse the motion of objects.

Activity 3 — Paths of ships

Use technology to create a simulation of this situation.

An oil-tanker, *Oil*, travelling at a constant speed has a position t hours after 12:00 given by the vector equation: $\begin{pmatrix} x \\ y \end{pmatrix} = \begin{pmatrix} 0 \\ 28 \end{pmatrix} + t \begin{pmatrix} 6 \\ -8 \end{pmatrix}$

where the vector $\begin{pmatrix} 1 \\ 0 \end{pmatrix}$ represents a displacement of 1 km due east and the vector $\begin{pmatrix} 0 \\ 1 \end{pmatrix}$ represents a displacement of 1 km due north. The port of *ORT* is situated at the origin.

a) Show the position of *Oil* at 13:00.

b) Draw a diagram to show the path of *Oil* and the port *ORT*.

c) Create a graph showing the distance between *Oil* and *ORT*. Deduce that a Cartesian equation for the path of *Oil* is $4x + 3y = 84$.

d) At what time will the distance between *Oil* and the port *ORT* reach its minimum value?

Another ship, the cargo-vessel *Cargy* is stationary, with position vector $\begin{pmatrix} 18 \\ 4 \end{pmatrix}$ km.

e) Show that *Oil* will collide with *Cargy* and find the time of the collision.

To avoid the collision, *Cargy* starts to move at 13:00, with velocity vector $\begin{pmatrix} 5 \\ 12 \end{pmatrix}$ km/h. The position of the *Cargy* is given by

$\begin{pmatrix} x \\ y \end{pmatrix} = \begin{pmatrix} 13 \\ -8 \end{pmatrix} + t \begin{pmatrix} 5 \\ 12 \end{pmatrix}$ for $t \geq 1$.

f) At 15:00 *Cargy* stops again. How far apart are the two ships at this time?

g) Find the distances between the port *ORT* and each of the ships at 15:00.

⌬ INTERNATIONAL MATHEMATICS

In different places the notation for vectors can be very different: the vector $\begin{pmatrix} 1 \\ 2 \end{pmatrix}$ is often represented by (1, 2), <1, 2>, [1, 2] or even $\mathbf{i} + 2\mathbf{j}$ where $\mathbf{i}$ and $\mathbf{j}$ are unit vectors in the positive horizontal and vertical directions. In general, the context of the problem makes the notation clear but it is essential that the notation is used consistently!

🌐 GLOBAL CONTEXT
Scientific and technical innovation

👤 ATL SKILLS
Thinking
Consider ideas from multiple perspectives.

TOPIC 2

Probability trees

Most of the probability theory you have learned so far is based on the probabilities of independent events. This means that the outcome of one event is not affected by the outcome of the other event in any way. Now you will look at the probabilities of dependent events, that is, the outcome of one event is affected by the outcome of the other. The probability that an event will occur when another event is known to have occurred is *conditional probability*.

For example, look at the probability of drawing two queens in a standard pack of playing cards, given that a pack contains 52 cards of which four are queens. First you calculate the chance of drawing a queen (4 out of 52), then calculate the probability of drawing a second queen, which is less likely now as one has already been drawn (3 of the 51 cards left are queens). So the conditional probability of drawing two queens is $\left(\dfrac{4}{52} \right) \times \left(\dfrac{3}{51} \right) = \dfrac{1}{221}$.

The use of diagrams to represent a situation is a very effective technique. It can help you to solve mathematical word problems as it enables you to visualize the situation and determine the necessary steps to find a solution.

In the next activity, you will be constructing probability trees as a visual way to represent the outcomes and conditional probabilities associated with games and medical testing.

Activity 4   Probability trees game

 INTERNATIONAL MATHEMATICS   "Two-up" is a game played in Australia, in which two coins (traditionally old pennies) are thrown into the air and players gamble on whether the coins will land with two heads up, two tails up, or one of each. It was played during the First World War by the Australian soldiers and is now played on Anzac Day in some pubs and clubs throughout Australia in memory of those soldiers.

STEP 1   Construct a probability tree to represent the possible outcomes and probabilities for tossing the two coins in Two-up.

STEP 2
a) **Use** the tree to calculate the probability of scoring two heads.

b) **Use** the tree to calculate the probability of scoring one of each.

c) What mathematical operation do you use when determining the probabilities on the same branch of the probability tree?

d) What mathematical operation do you use when determining the final probabilities on different branches of the probability tree?

TIP

Remember that if you are each taking just a few turns at the game it may not give the desired probabilities. The more games that you play, the closer the probability will get to its theoretical probability.

🌐 **GLOBAL CONTEXTS**
Scientific and technical innovation

🧠 **ATL SKILLS**
Thinking
Revise understanding based on new information and evidence.

Probability has many applications and most of them are very different from trying to predict the outcome of games. Industry, commerce, financial services and medicine all use probability as a valuable tool. In the next activity, you will investigate some very important research.

 # Activity 5   Conditional probability and medical test results

In mid-2012, the US Food and Drug Administration (FDA) approved a new over-the-counter HIV test. It only requires a saliva sample and will produce results within 40 minutes. When tests like this become available to the public, everyone asks how accurate such a test should be. Most medical tests have a slight margin for error. It is important to understand how significant these errors can be when taking into account the size of a population.

These are research test results.

The sensitivity test—92 per cent of people who are HIV positive received a positive test result. Based on this result, what fraction of the population that was tested received a false positive test (a positive result even though they are really HIV negative)?

The specificity test—99.98 per cent of people who were HIV negative received a negative test result. Based on this result, what fraction of the population that was tested received a false negative test?

 **a)** Research the current population of your country.

b) Research the approximate number of people who are HIV positive in your country.

STEP 2 Set up a probability tree to represent all possible outcomes when taking this test.

STEP 3 Use the probability tree to answer these questions.

a) What is the probability that the test is correct?

b) Given that a person in your country receives a positive test result, what is the probability that the person is actually HIV negative (received a false positive)?

c) Given that a person in your country receives a negative test result, what is the probability that the person is actually HIV positive (received a false negative)?

REFLECTION

a) Which result do you think it is more important to minimize, a false positive or a false negative? Justify your answer.

b) Do you think that these probabilities are sufficiently minimized for this test?

c) Can you see such a test reducing the transmission rate of HIV? Explain .

🌐 **GLOBAL CONTEXTS**
Identities and relationships

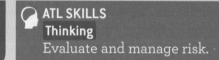 **ATL SKILLS**
Thinking
Evaluate and manage risk.

TOPIC 3

Misrepresentation

A chapter on representation wouldn't be complete without a section on misrepresentation. For some people, simply reading or hearing numbers and data can be very convincing. In this section, you will see how data can be used to lead or mislead people.

Standard deviation is a measure of spread, it tells you how far from the mean the data values vary. The next activity shows how the standard deviation can be used in statistical surveys.

 Activity 6   **Standard deviation**

Look at the graph below and then answer the questions.

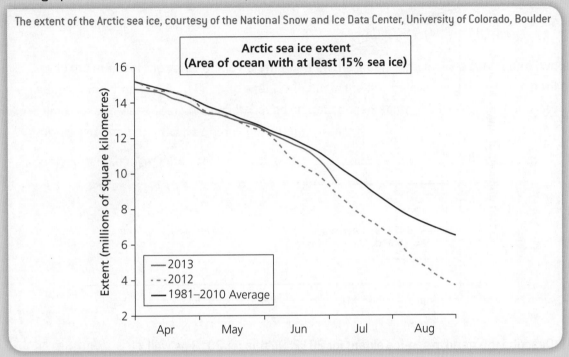

The extent of the Arctic sea ice, courtesy of the National Snow and Ice Data Center, University of Colorado, Boulder

a) [Describe] what is happening to the level of sea ice in the months shown.

b) [Explain] how you could use this graph to argue that global warming has been having an increasingly detrimental effect on sea ice coverage since the 1980s. Is it possible to use the graph to argue that the effects are not due to global warming?

c) Research standard deviation.

- What does it measure?
- How is it calculated?
- What does it mean to be "within one standard deviation of the mean" or "within two standard deviations of the mean"?

d) Are the trends in the graph you have been looking at out of the ordinary? [Explain].

This graph represents the same data, but now ±2 standard deviations are also shown (the grey area).

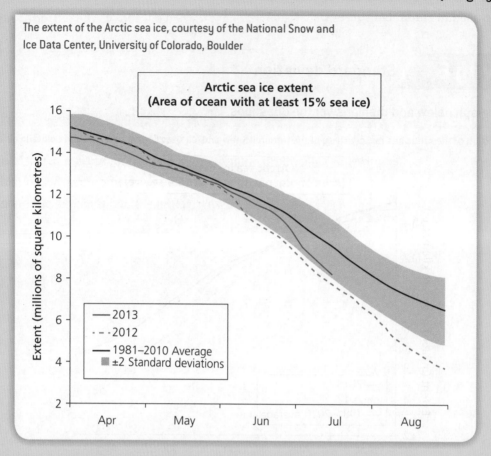

The extent of the Arctic sea ice, courtesy of the National Snow and Ice Data Center, University of Colorado, Boulder

**Arctic sea ice extent
(Area of ocean with at least 15% sea ice)**

Legend:
— 2013
- - - 2012
— 1981–2010 Average
■ ±2 Standard deviations

Y-axis: Extent (millions of square kilometres)
X-axis: Apr, May, Jun, Jul, Aug

e) Describe the trend in sea-ice extent for 2012. Do this for 2013 as well.

f) **Explain** how having the standard deviation represented on the graph affects your interpretation of whether or not the sea-ice changes in 2012 and 2013 are a cause for concern.

g) What else could be done with the graphs in order to misrepresent the data? Describe two options that produce two very different conclusions that could be drawn from this "real" data.

🌐 **GLOBAL CONTEXTS**
Globalization and sustainability

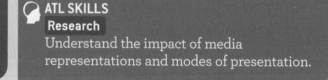

ATL SKILLS
Research
Understand the impact of media representations and modes of presentation.

Because standard deviation measures how spread out data are, it is often referred to as a measure of dispersion. In previous mathematics courses, you may have studied measures of central tendency, mean, median and mode—measures that attempt to describe the middle of a data set. While the measures of central tendency may not be difficult to calculate, knowing when to use each appropriately is an important part of determining whether or not data is being misrepresented.

 **Activity 7**   **Measures of central tendency**

a) Define the three statistical measures of central tendency.

b) **Describe** and discuss situations in which it is appropriate to use the mode to represent a set of data. When is it inappropriate? Give an example of a data set for which the mode is the most appropriate choice to represent it. **Explain** why this is the case.

c) **Describe** and discuss situations in which it is appropriate to use the median to represent a set of data. When is it inappropriate? Give an example of a data set for which the median is the most appropriate choice to represent it. **Explain** why this is the case.

d) **Describe** and discuss situations in which it is appropriate to use the mean to represent a set of data. When is it inappropriate? Give an example of a data set for which the mean is the most appropriate choice to represent it. **Explain** why this is the case.

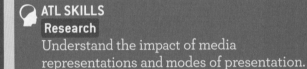

🌐 **GLOBAL CONTEXTS**
Personal and cultural expression

🗣 **ATL SKILLS**
Research
Understand the impact of media representations and modes of presentation.

Reflection

a) What does it mean to be "media literate"? How can mathematics help you to become more media literate?

b) Many people would say that all newspaper or television reports contain some amount of bias. Can data and statistics be biased? **Explain**.

c) Find an example in the media that contains a misrepresentation of mathematical data. **Explain** how the data has been manipulated.

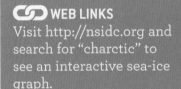

🔗 **WEB LINKS**
Visit http://nsidc.org and search for "charctic" to see an interactive sea-ice graph.

Summary

In this chapter, you have explored different ways of representing mathematical ideas and seen how they each have their own advantages and disadvantages. Despite this, these different representations all have their usefulness and applications in a wide range of fields. You have also seen that how you represent a problem can greatly simplify your ability to analyse and solve it. However, there is always a danger that anything that can be represented can also be misrepresented. While numbers and data can be very convincing, the way in which they are represented can also affect the message you are trying to convey.

Simplification

The process of reducing to a less complicated form

TOPIC 1 Simplifying algebraic and numeric expressions
- Is simpler always better?

TOPIC 2 Simplifying through formulas
- If you could find an easy way out, would you take it?

TOPIC 3 Simplifying a problem
- Do simpler problems have simpler solutions?

SKILLS

ATL

✓ Apply skills and knowledge in unfamiliar situations.

✓ Make connections between subject groups and disciplines.

✓ Apply existing knowledge to generate new ideas, products or processes.

✓ Analyse complex concepts and projects into their constituent parts and synthesize them to create new understanding.

✓ Process data and report results.

✓ Revise understanding based on new information or evidence.

Algebra

✓ Identify terms in arithmetic and geometric sequences.

✓ Simplify using the laws of logarithms and exponents.

✓ Factorise quadratic expressions and solve quadratic equations.

✓ Divide polynomials.

✓ Factorise higher degree polynomials and solve polynomial equations.

✓ Graph data and find the gradient of a line.

✓ Solve exponential equations.

✓ Use a linear programming model to help analyse a situation and make business decisions.

Geometry and trigonometry

✓ Calculate metric relations in right-angled triangles.

✓ Use basic trigonometric ratios and/or the sine rule to derive formulas to find the heights of objects.

**OTHER
RELATED
CONCEPTS**

**Equivalence Justification Model Pattern
Representation**

GLOSSARY

Daughter isotope the product which remains after an original isotope has undergone radioactive decay.

COMMAND TERMS

Deduce reach a conclusion from the information given.

Find obtain an answer showing relevant stages in the working.

Identify provide an answer from a number of possibilities. Recognize and state briefly a distinguishing fact or feature.

Solve obtain the answer(s) using algebraic and/or numerical and/or graphical methods.

State give a specific name, value or other brief answer without explanation or calculation.

Introducing simplification

In mathematics, you often see the instruction to "simplify" or "write in simplest terms". It probably first began with fractions and then steadily moved into much of what you do in algebra, from exponents to polynomials to radicals. Every new topic brings with it a need to reduce expressions to a less complicated form–simpler is better. Some of the most useful results you have studied in mathematics are rather simple but yet very powerful. For example, Pythagoras' theorem can be presented in a very simple form visually: the red square has the same area as the green and purple squares together.

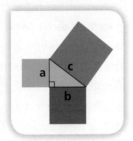

You already know that expressions can be presented in different equivalent forms. Sets of rules, such as the order of operations, enable you to convert one form of an expression into a simpler form. Think how you would simplify the expression:

$$3(5 - 2) + 4^2 - 12(3) \div 4$$

and how much more convenient the simpler form would be.

Simplifying more complicated expressions may require more complicated rules, such as the laws of exponents, logarithms or radicals. Ultimately, the goal is to be able to use the rules to simplify expressions so that they can then be applied in other circumstances, such as solving equations.

Simplification is also an important strategy in solving problems. If you can't find the answer to a question, try solving a simpler version first. By doing a little work beforehand, you may even be able to develop formulas and strategies that will simplify some of those problems. For example, you can find the area of a trapezium by breaking it up into simpler shapes with areas that you can calculate easily:

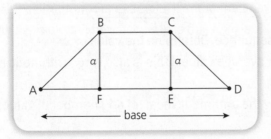

However, there is a formula for finding the area of a trapezium. Once you know the formula, you can easily find the area of any trapezium in the future.

In this chapter, you will explore variations on the concept of "simplification": writing in simplest form, solving a simpler problem and using formulas to simplify the problem-solving process.

CHAPTER LINKS
In Chapter 7 on generalization, students apply the strategies in Polya's problem-solving method.

WEB LINKS
For the derivation of the formula for the area of a trapezium, see www.mathopenref.com and search for "trapezium".

Simplifying algebraic and numeric expressions

In the following activities, you will consider some ideas with which you may already be familiar, but you will think about them in different ways. In previous mathematics courses you have already worked with the *arithmetic mean* of a set of numbers. This is an average value. Another type of average value is called the *geometric mean*. There are several ways to calculate the geometric mean. You will use one of them in the next activity. Once you have completed this method, you will use the laws of logarithms to simplify the formula to arrive at a much simpler version.

 Activity 1   The geometric mean

Did you know that, when you are looking at a work of art, there is supposedly an "ideal location" where you should stand?

Interestingly, you can use the geometric mean to calculate that location mathematically. Have a closer look at the two mean values.

STEP 1
a) The terms 2, x, 8 form an arithmetic sequence. Determine the value of x.
b) [State] how this relates to the mean or average value. (This is also why "arithmetic mean" is an appropriate name.)
c) Write down the formula for calculating the arithmetic mean of any number of values. Use appropriate mathematical notation.
d) Use the formula to find the arithmetic mean of 4, 9 and 14.

STEP 2
a) The terms 2, y, 8 form a geometric sequence. Determine the value of y, given that $y > 0$. This value is the geometric mean of 2 and 8.
b) [Find] the geometric mean of 5 and 180.
c) Can you find a simple way to calculate the geometric mean of two values? If so, write it down. Use correct mathematical notation.

STEP 3 Here is one way to calculate the geometric mean of three numbers.

i) Take the logarithm (base 10) of each value.

ii) Find the arithmetic mean of these logarithmic values by adding them together and dividing by 3.

iii) Raise 10 to the power you have calculated in ii.

Use this method to find the geometric mean of 4, 9 and 15.

> **TIP**
>
> When no base is specified for a logarithm, it is assumed to be base 10.

STEP 4 If you wrote all of these steps as one expression, you would have

$$10^{\frac{(\log 4 + \log 9 + \log 15)}{3}}.$$

In fact, you can calculate the geometric mean of *any* number of values in a similar way.

$$10^{\frac{\left(\log x_1 + \log x_2 + \log x_3 + \cdots + \log x_n\right)}{n}}$$

a) Use the laws of logarithms to simplify the above expression. You should arrive at another method for calculating the geometric mean (that does not require logarithms). Explain this simpler method in words.

b) Use the laws of logarithms in a different order to arrive at the same simplified form.

> **INTERDISCIPLINARY LINKS** Logarithms are used extensively in science. In chemistry, students learn to use logarithms to calculate the pH of a substance. In physics, sound intensity is measured in decibels, which also involves a logarithmic scale.

c) Simplify each of these logarithmic expressions.

i) $\log x + \log y + \log z^4$

ii) $\log_5 \sqrt{x} + \log_5 y^2 + \log_5 z^{-2}$

iii) $2\log_3 x + 3\log_3 y^4 - 4\log_3 z$

iv) $\frac{1}{3}\ln x - 4\ln y + \frac{1}{2}\ln z$

v) $2\left(\log_6 x^3 - \frac{1}{4}\log_6 y - \log_3 z^2\right)$

How does all of this relate to looking at art in a museum? The ideal place to stand, when looking at a painting, is at the geometric mean of the distance from your eye to the top of the painting and the distance from your eye to the bottom of the painting. When you are standing at this horizontal distance from the picture, you maximize the angle to the top and bottom of the painting and supposedly get the best view. This is known as the "Regiomontanus' Hanging Picture Problem".

> **WEB LINKS**
> Visit http://hom.wikidot.com/regiomontanus to learn more about The Regiomontanus' Hanging Picture Problem.

GLOBAL CONTEXTS
Personal and cultural expression

ATL SKILLS
Transfer
Apply skills and knowledge in unfamiliar situations.

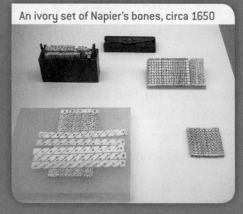
Activity 2   Dimensional analysis

Physicists use dimensional analysis to verify that appropriate units result when using a formula or equation. When units on both sides of an equation are equal, there is no guarantee that the equation is correct. However, when units on either side are not equal to each other, then there is definitely something wrong with the formula or equation. Working with the units while performing calculations is a good way to verify that the result is appropriate.

> **TIP**
>
> It is worth noting that science and maths often use the same symbols but with different meanings.

a) Research the definition of one newton in terms of other units.

b) The force of gravity (F) between any two objects of mass m_1 and m_2, separated by a distance r is given by:

$$F = \frac{Gm_1m_2}{r^2}$$

where the masses are measured in kilograms, the distance between them is measured in metres and the force is measured in newtons.

The universal gravitation constant, G, is measured in $m^3kg^{-1}s^{-2}$. Use the laws of exponents to simplify the units on the right-hand side to show that the units on both sides of the equation are the same.

c) The time (T, measured in seconds) it takes for a pendulum to swing back and forth once is related to both the length of the pendulum (L, measured in metres) and the acceleration due to gravity (g, measured in ms^{-2}):

$$T = 2\pi\sqrt{\frac{L}{g}}$$

Use the laws of exponents to simplify the right-hand side to show that the units are the same on both sides of the equation.

d) When an object accelerates from an initial velocity (V_i, measured in m/s) to a final velocity (V_f, measured in m/s) over a distance (d measured in metres), its final velocity is given by:

$$V_f = (V_i^2 + 2ad)^{\frac{1}{2}}$$

where a is the acceleration (measured in m/s²). Use the laws of exponents to simplify the right-hand side to show that the units are the same on both sides of the equation.

e) i) The electric field ($\mathcal{E}$, measured in newtons per coulomb, N/C) due to a line of charge is given by:

$$\mathcal{E} = \frac{2k\lambda}{r}$$

where λ is the charge per unit of length (measured in C/m), r is the distance (in metres) and k is a constant (measured in Nm²/C²). Use the laws of exponents to simplify the right-hand side to show that the units are the same on both sides of the equation.

ii) The electric field due to a ring of charge is given by:

$$\mathcal{E} = \frac{kqz}{\left(z^2 + R^2\right)^{\frac{3}{2}}}$$

where q is measured in coulombs (C), and z and R are both measured in metres. Use the laws of exponents to simplify the right-hand side to show that the units are the same on both sides of the equation.

REFLECTION

a) Explain how the rules for simplifying logarithms (Activity 1) and exponents (Activity 2) are similar.

b) Describe what you can do if you don't remember the rules for simplifying an exponential expression.

c) In which contexts is 24 simpler to use than $2^3 \times 3$? In which contexts is $2^3 \times 3$ simpler to use than 24? Explain.

 **GLOBAL CONTEXTS**
Scientific and technical innovation

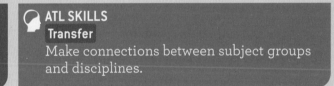 **ATL SKILLS**
Transfer
Make connections between subject groups and disciplines.

TOPIC 2

Simplifying through formulas

Formulas are an integral part of many areas of knowledge, especially within sciences or mathematics. Formulas are also used in areas such as economics, geography and statistics. However, simply knowing a formula without knowing why it is appropriate, in fact, restricts your ability to solve problems efficiently. To simplify problem-solving in both mathematics and science, you need to know why formulas work and how they can be applied correctly. In this topic, you will see how to apply your knowledge of right-angled triangles to discover new formulas that can simplify future problems with these shapes.

Suppose you were asked to find the values of h, m and n in this diagram.

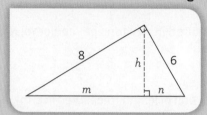

STEP 1   **a)**   State the different methods that you could use to solve the problem. Be specific about the mathematics involved in each method.

       b)   Use the methods you listed to solve the problem. Which method was the simplest?

One of the methods that you may use to solve this problem involves similar triangles. You can use the properties of similar triangles to develop relationships called metric relations.

STEP 2   **a)**   Start with a triangle.

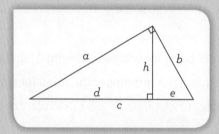

Determine which triangles are similar. Work out proportion statements that relate each set of lengths.

       i) a, b, c and h **ii)** d, e and h **iii)** a, c and d **iv)** b, c and e

       b)   Rewrite each of the proportion statements so that there is no division, only multiplication. These are the metric relations in a right-angled triangle.

STEP 3   **a)**   Use these metric relations to solve the original problem. Would knowing these beforehand have simplified the problem?

       b)   Find the values of x and y in this diagram.

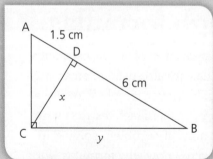

REFLECTION

a) These metric relations are specific to right-angled triangles. How does that influence your opinion of their usefulness?

b) The relationship between a, b, c and h is not surprising when you take a closer look at it. What else is that relation based on, apart from similar triangles?

> **ATL SKILLS**
> **Thinking**
> Apply existing knowledge to generate new ideas, products or processes.

👤 Activity 4   Deriving formulas for other problems involving right-angled triangles

STEP 1 Work through this problem.

Imagine you are in Melbourne, Australia and see the Rialto Tower at an angle of elevation of 47° from a point on a level section of road. When you move 90 m closer to the tower, along the same level section of the road, the angle of elevation is 60°. What is the height of the Rialto tower, to the nearest metre?

STEP 2 This type of trigonometry question involves multiple steps, and can take some time to complete. If you take the time to derive a formula, you can use it to simplify the process and make these questions easier to answer when you meet them again.

To derive the formula, you need to use variables instead of numbers. Use the variables shown in this diagram, and follow through exactly the same process as you did for step 1. **Deduce** a formula to find the height in terms of d and θ_1 and θ_2.

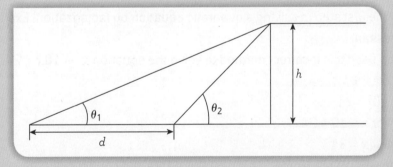

Using your formula check your answer to step 1.

REFLECTION

Is it worth taking the time to derive the formula to simplify future similar problems? Justify why or why not.

How successful would a student be in mathematics if all they knew were formulas? What would they be lacking?

GLOBAL CONTEXTS
Scientific and technical innovation

ATL SKILLS
Thinking
Apply existing knowledge to generate new ideas, products or processes.

TOPIC 3

Simplifying a problem

You already know how to solve linear equations such as $3x + 2 = 7$. You have practised them so much that solving them has almost become automatic it is so simple. More recently, you began solving quadratic equations of the form $ax^2 + bx + c = 0$. While there are various methods for solving these, one of them in particular requires your skills with linear equations, thereby simplifying the problem. You will start with that method and find out if you can use the same approach with other polynomial equations.

Activity 5   **Polynomial equations**

STEP 1   **Quadratic equations**

a) Explain how you know that an equation is a quadratic.

b) **State** what it means if an expression is a factor of a polynomial?

c) **Solve** these equations by factorizing.

  i) $x^2 - 5x - 6 = 0$ ii) $2x^2 + x - 15 = 0$

  iii) $x^2 + 24 = 14x$ iv) $4x^2 - 3x = 27$

d) **State** the first step in solving a quadratic equation by factorization. Explain why it is necessary.

e) Gerardo used this incorrect method to solve the equation $x^2 - 10x = 24$.

$x^2 - 10x = 24$

$x(x - 10) = 24$

$x = 24$ or $x - 10 = 24$

$x = 24$ or $x = 34$

Explain where Gerardo made his mistake. **State** the correct solutions.

f) Describe how you simplified quadratic equations so you could solve them.

Other polynomial equations

Suppose you wanted to solve $x^3 - 2x^2 - 5x + 6 = 0$.

a) Divide $x^3 - 2x^2 - 5x + 6$ by $x + 1$. Is $(x + 1)$ one of its linear factors? Explain.

b) Divide $x^3 - 2x^2 - 5x + 6$ by $x - 1$. Is $(x - 1)$ one of its linear factors? Explain.

c) What is the quotient in b)? Can you factor it?

d) Write $x^3 - 2x^2 - 5x + 6$ as a product of its linear factors.

e) Solve $x^3 - 2x^2 - 5x + 6 = 0$.

f) Explain how the original problem was simplified so that its solution could be found.

g) Explain why you still need to manipulate the equation so that one side is equal to zero before factorizing.

h) Solve these equations by factorizing. You may need to use long division in some cases.

i) $2y^3 = 18y$ ii) $x^3 + 2x^2 - 4x - 8 = 0$ iii) $m^3 - 3m^2 = 16m - 48$

ATL SKILLS

Thinking

Analyse complex concepts and projects into their constituent parts and synthesize them to create new understanding.

Activity 6   **Dating the Earth**

Working out the age of a tree is easy—once you have cut it down. You just count the growth rings. But what if you don't want to cut down the tree? What if you want to find the age of something that doesn't have growth rings?

You might use carbon-14 dating, which is based on how long it takes carbon-14 to decay to carbon-12. Carbon-14 and carbon-12 are both isotopes of carbon.

Rings in the cross-section of a tree trunk

Although carbon dating is very useful, it is limited to objects that are less than 50 000 years old. So how could you date objects that are even older? How, for example, would you find the age of the Earth?

Scientists have found other naturally occurring isotopes that also decay. One of those is rubidium-87 (the parent isotope) which, over time, decays and becomes strontium-87 (the daughter isotope).

It can be shown that the amount of the **daughter isotope** formed, D, is related to the amount of parent isotope left, P, by the equation:

$$D = P(e^{\lambda t} - 1) + D_\circ$$

where $D_\circ$ is the initial amount of the daughter isotope present and λ is the decay constant for the parent isotope which, in this case, is 1.42×10^{-11}. The time t is measured in years.

This relationship is difficult to graph on a two-dimensional coordinate system because there are three variables. Both D and P change as time passes, as does the value of t. Because of this, scientists will often use a process called *linearization* to simplify their work. The next task should help you to find out why this applies here.

a) Explain how the above equation is similar to the gradient—intercept form of a linear function $(y = mx + b)$.

b) Which quantity, P or D, is the independent variable? Which is the dependent variable?

c) If a scientist plots D against P on the same coordinate plane, what part of the above equation represents the y-intercept? What part should be equal to the gradient of the line?

Now you will use the result in **c** to determine the age of the Earth.

A scientist has five samples of meteorites found on Earth. She measures their rubidium-87 content (P) and their strontium-87 content (D). Her results are recorded in this table.

Meteorite	Rubidium-87	Strontium-87
A	1.72	1.514
B	1.6	1.502
C	1.44	1.494
D	1.2	1.478
E	0.18	1.412

d) Using an appropriate scale, graph the data.

e) Draw the line of best fit and determine its gradient.

f) Use the gradient and your result in **c** to find the age of the meteorites and, thereby, the age of the Earth.

REFLECTION

a) Explain how graphing D against P simplified the problem of trying to graph the relationship given by $D = P(e^{\lambda t} - 1) + D_o$.

b) State any assumptions that you made in solving the problem.

c) Experimentally, it can be shown that the gravitational force (F) between any two bodies is inversely related to the square of the distance (r) between them, $F = \dfrac{Gm_1 m_2}{r^2}$. If you collected data for F and r, what would you graph in order to simplify the task of finding the equation that relates these two quantities?

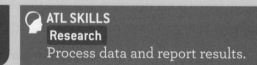

Systems of inequalities

The simplex method is used in linear programming to find the optimum value of a function that is subject to constraints (expressed as inequalities). The inequalities define a polygonal region and the solution is one of the vertices of this region. When expressed in graphical form, the polygonal region produced by the simplex method gives a very clear, simplified representation of all of this information. The best possible solution can be easily identified. The complexity of the problem increases with the addition of variables and constraints. Such problems are usually solved by means of a computer program.

🏛 MATHS THROUGH HISTORY

The simplex method was created by George Dantzig, a mathematician working for the US Air Force during the Second World War. He developed linear programming as an efficient method of allocating military resources and improving strategic planning. Since then, businesses have been using linear programming in an effort to optimize their use of resources and maximize profit.

Activity 7 — Allocating resources to maximize profit

A small company needs to decide how to allocate their resources to maximize profit.

A high-end suspension mountain bike manufacturer produces two different bike models in one workshop: a cross country trail bike (XC) and a gravity downhill trail bike (GD). Each type of bike must be processed through three stages: component manufacture, assembly and finishing (painting and polishing, and so on).

Each XC bike requires five hours' work in the component manufacture workshop and four hours in the assembly workshop.

Each GD bike requires ten hours in the component manufacture workshop and six hours in the assembly workshop.

Each bike takes one hour in the finishing workshop.

The component manufacture, assembly and finishing workshops are available for 165, 120 and 28 hours per week, respectively.

The profit contribution per bike is $600 for each XC bike and $750 for each GD bike.

STEP 1 — Assuming that the manufacturer wishes to maximize profits, determine how many bikes of each type should be produced per week.

a) Determine the inequalities to represent the objective function and the constraints.

b) Explain each inequality and how it affects the scenario.

c) Graph all inequalities by hand on graph paper and determine all of the corner points.

TIP

There are many GDCs and downloadable graphing software applications that will graph inequalities.

d) Use technology to help you visualize your graph.

    i) How many bikes of each type should be produced per week?

    ii) What is the maximum profit the manufacturer can make per week?

    iii) Use a spreadsheet with a 'solver' function to validate your answer.

    iv) Is the equipment in the workshop being fully utilized? If not, how long is any of the machinery not in use?

    v) Explain how you would represent the profit equation graphically. Find a possible profit equation.

    vi) Explain how you can use the profit equation graphically to validate that your answer is correct.

> **TIP**
>
> Excel solver is available on any Excel program and can be used to solve complex linear programming models.

STEP 2 **Adapting the linear programming model**

Use a GDC or graphing software to complete this part of the activity.

Scenario one

There have been changes in regulations regarding length of shifts and changeover between them. In addition, there are new requirements for maintenance of equipment. As a result, the component manufacturing workshop is only available for 150 hours per week. How many bikes of each type should now be produced per week? How much profit will they make?

Explain any assumptions you have made.

Explain the problem-solving process you used to solve this question.

Scenario two

Assume the component manufacturer equipment is still only available for 150 hours per week. An important customer owns a chain of shops and has requested a standard order of six downhill gravity bikes per week.

Explain why you would (or would not) accept this order. Justify your answer. Use quantitative data.

Scenario three

Assume the component workshop is still only available for 150 hours per week and you have no restrictions on the number of each type of bike produced. You could buy a new piece of spray-painting equipment so that the finishing time would be reduced to three-quarters of an hour for each bike. To pay for the equipment, your profit will be reduced by 20 dollars per bike. Would you buy the equipment?

Explain whether or not you would buy the equipment. Justify your answer. Use quantitative data.

REFLECTION

a) Explain why the vertices of the polygon defined by the inequalities (the corner points) are so important in linear programming.

b) [Identify] the limitations that you experience with creating a linear programming model.

c) Research three examples in which linear programming is used in decision-making. List how many possible variables and constraints these models can have. Would it be possible to solve these models on ordinary graph paper? Explain why or why not.

GLOBAL CONTEXTS
Scientific and technical innovation

ATL SKILLS
Thinking
Revise understanding based on new information or evidence.

Summary

In this chapter, you have looked at how you can use predetermined rules to simplify both algebraic and numeric expressions. Problems can sometimes be simplified by developing formulas, using mathematics that you already know. A common theme in both of these aspects of simplification is that you must not only understand what the rules are but also how to apply them correctly. You have also seen that sometimes the problems themselves can be simplified, making finding a solution easier. Simplification in mathematics is not therefore just about reducing expressions to a less complicated form, it also refers to the problems you solve and the strategies you use to solve them.

INQUIRY QUESTIONS	**TOPIC 1** Special points and lines in 2D shapes ▪ **What makes a point special?** **TOPIC 2** Mathematics and art ▪ **Can creativity be planned?** **TOPIC 3** Volume and surface area of 3D shapes ▪ **How can design lead to a better use of resources?**

SKILLS

ATL

✓ Listen actively to other perspectives and ideas.

✓ Develop new skills, techniques and strategies for effective learning.

✓ Create novel solutions to authentic problems.

Number

✓ Calculate the percentage increase and decrease of surface area and volume to determine the best configuration.

Algebra

✓ Solve systems of linear equations.

✓ Solve problems involving geometric sequences and series.

Geometry and trigonometry

✓ Recognize lines of symmetry of shapes.

✓ Recognize the properties of polygons.

✓ Calculate the coordinates of midpoints and the distance between points.

✓ Use properties of triangles to find the location of their special points (centres).

✓ Explore the characteristics of the platonic solids.

✓ Analyse plane sections of solids.

✓ Calculate the surface area and volume of rectangular prisms and cylinders.

✓ Calculate perimeters and areas of plane figures.

✓ Use Pythagoras' theorem to find unknown side lengths.

OTHER RELATED CONCEPTS

Quantity   Model   Representation

GLOSSARY

Configuration an arrangement of parts or elements in a particular form, figure, or combination.

Platonic solid a 3D shape where each face is the same regular polygon and the same number of polygons meet at each vertex (corner).

COMMAND TERMS

Calculate determine the amount or number of something mathematically.

Describe give a detailed account or picture of a situation, event, pattern or process.

Explain give a detailed account including reasons or causes.

Show allow, cause to be visible.

Introducing space

MATHS THROUGH HISTORY

It is said that one day while Rene Descartes was lying in bed staring at the ceiling, a fly buzzed in and landed on it. Descartes thought, "How can I describe the location of the fly on the ceiling to someone who isn't in the room, so that they know exactly where the fly is?" He thought of the ceiling as being divided into four parts, or quadrants. The location of the fly could now be uniquely determined by two numbers, an ordered pair. The first number indicates the position of the fly to the right or left of the origin, and the second number indicates its position above or below the origin. This is the Cartesian plane that you are familiar with, named after Rene Descartes, a French mathematician who lived from 1597 to 1650.

Geometry is grasping space ... that space in which the child lives, breathes and moves. The space that the child must learn to know, explore, and conquer, in order to live, breathe and move better in it.

Hans Freudenthal

How would you describe the position of a fly buzzing around in your room, not just on the ceiling, but at any point in the room? You would need to add another axis to the x–y plane, to identify the unique position of the fly in your three-dimensional space. Any point in space has the three unique dimensions of length (longitude), width (latitude) and depth (or altitude), relative to some given origin.

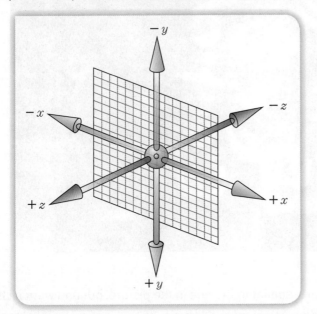

Having established that the world is three-dimensional, you also need to remember that you move through time. Scientists consider time to be the fourth dimension. You can locate a unique 3D position in space, together with a particular instant in time. This means that nobody can be in two different locations at the same time.

INTERDISCIPLINARY LINKS

It is thought that the great artist Pablo Picasso painted his *Portrait of Dora Maar* in an attempt to show how a three-dimensional object might look like to a fourth-dimensional being. You can search for the painting online.

Special points and lines in 2D shapes

In this topic you will make a cardboard triangle and try to balance it on the tip of a pencil. You will need to experiment with different points inside the triangle. The point where you can balance it is the centroid of the triangle, or the barycentre. It is the centre of gravity of the triangle.

 WEB LINKS
Go to www.mathopenref.com and search for "triangle centroid".

Activity 1 — Get to the point!

Suppose that you plan to build a mobile sculpture, similar to the one in the picture, but you want all the shapes to hang horizontally. You need to decide how to hang each shape, so that it stays parallel to the floor. You need to find the centroid of each shape.

STEP 1 Find the centroid of each of these shapes. Make cardboard cut-outs of the shapes, if it helps.

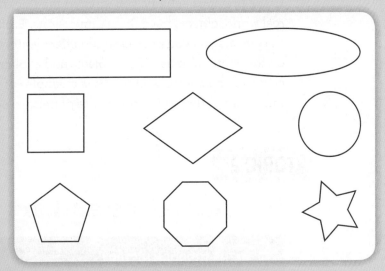

In each case, **describe** how you found the centroid. You could use cardboard models to test your theory.

STEP 2 Now consider different triangles: equilateral, isosceles and scalene. Investigate the location of the centroid in each of them. Write a report explaining how you found their centroids and the properties of these triangles that you have identified.

STEP 3 Consider the triangle defined by the points A(1, 4), B(7, 2) and C(5, 8) in the Cartesian plane.
  a) Find the midpoints M, N and P of the segments AB, AC and BC, respectively.
  b) Find Cartesian equations for the lines MC, NB and PA.
  c) **Show** that the lines MC, NB and PA have a common point G and find its coordinates.
  d) **Show** that AG : GP = BG : GN = CG : GM and state the value of these ratios.
  e) **Describe** the relation between the location of G and the centroid of triangle ABC. Does the relation hold for other triangles? Write down your conclusions.

STEP 4 Construct your own mobile structure. Use at least six different shapes.

Special effects in movies often depended on finding the centroid of an object. For example, they suspend the spaceship from its centroid to make it seem to fly without wobbling! This same principle enabled people to fake credible UFO sightings, as long as the person holding the model had a steady hand!

UFO

🌐 **GLOBAL CONTEXTS**
Personal and cultural expression

💭 **ATL SKILLS**
Thinking
Propose and evaluate a variety of solutions.

Reflection

In the previous activity, you ignored the thickness of the object. In the real world, objects have three dimensions. **Explain** how the thickness of the objects you considered would affect your results. Do some research on the centre of mass of 3D objects and explain the relation between centroid and centre of mass. Give examples of situations in which it is important to determine the centre of mass to solve a real-life problem.

TOPIC 2

Mathematics and art

> 🏛 **MATHS THROUGH HISTORY**
>
> Historically, mathematics and art have always enjoyed a very close relationship. Beautiful symmetrical, decorative patterns can be seen in mosques; perspective has been used for hundreds of years, in drawings and paintings, to create the illusion of 3D space. The golden ratio of length to height has been known since ancient times and was used in the creation of famous structures such as the Parthenon in Greece and the Pyramids of Egypt. More recently, mathematics can be seen to influence art in the form of Cubism.

A polyhedron is a solid with flat surfaces. Five polyhedra, known as the **Platonic solids**, are different from all the other solids. The aesthetic beauty and symmetry of the Platonic solids have fascinated mathematicians for thousands of years.

 Activity 2 **Properties of 3D objects and impossible objects**

In this activity, you will look at platonic solids, a subject that has been the favourite subject of many mathematicians.

> **WEB LINKS** Visit http://www.coolmath4kids.com and look at the "polyhedra" section.
> To view virtual reality polyhedra, go to www.georgehart.com, then follow the links to the "Encyclopedia of Polyhedra".

STEP 1 The most well-known platonic solid is probably the cube. This is the shape of regular dice.

The cube is a special platonic solid. It is the only one with faces that have an even number of sides. Its faces are squares. You can easily make a model of a cube from cardboard. All you need is a net like this one.

In fact there are exactly 11 possible nets for a cube. Draw all of them.

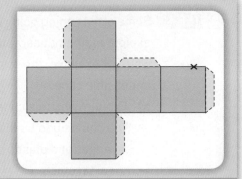

STEP 2 It is possible to make a cube out of three congruent square-based pyramids, as shown in the picture.

- Draw a net that allows you to produce a model like the one in the picture.
- What does this model tell you about the relation between the volume of the cube and the volume of each pyramid?
- Modify your net to produce three congruent pyramids that can be assembled to form a cuboid (rectangular prism) with dimensions 4 cm by 5 cm by 6 cm. State the volume of each pyramid.
- **Explain** how to obtain the volume of a rectangular pyramid with base of dimensions $a \times b$ and height h.

STEP 3 Cubes are often used in sculptures. Porto, in Portugal, even has a cube as one of its landmarks.

As this monument is near a river, whenever floods occur different sections of the cube can be observed. The intersection of the cube with a plane (the surface of the river), T, is always either an equilateral triangle or a regular hexagon.

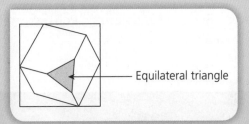
— Equilateral triangle

Porto fountain
© adamfrunski (via Flickr)

a) Make model cubes from a material that can be easily cut (for example, potato). Cut off corners of these models to show the following sections:
- equilateral triangle
- isosceles triangle
- scalene triangle.
In each case, **describe** how these sections can be produced.

b) Use models of cubes to explore all possible sections that can be produced on the cube. Investigate which regular polygons can be obtained as sections of a cube. Make a table showing sketches of the sections, their names and how they can be obtained.

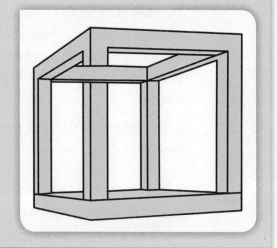

STEP 4 Some artists enjoy going beyond the real world and explore mathematical concepts in a very creative way. For example, the sculpture here shows an impossible cube.

a) **Explain** why this cube is an impossible object. You could try to build a model, using toothpicks as edges and plastic modelling clay to hold them together.

b) Research and find examples of other sculptures that are impossible objects. Choose an example and **explain** how the artist designed it.

🌐 **GLOBAL CONTEXTS**
Personal and cultural expression

🗣 **ATL SKILLS**
Self-management
Develop new skills, techniques and strategies for effective learning.

Fractal geometry

A fractal is a geometric object created by repeating a pattern many, many times on an ever-decreasing scale. If you look at the outline of the fractal below, you can see ever smaller repetitions of the larger picture all along its boundaries. If you were to zoom in on the boundary, you would find ever smaller ones, seeming to go on for ever. Therefore, although the fractal seems simple and orderly, it has complex properties that make it chaotic as well. Fractals are simple and complex at the same time, exhibiting chaos and order simultaneously. The most famous fractal is the Mandelbrot set, discovered by Benoit Mandelbrot, a Polish-born French and American mathematician.

Figure 15.1 The Madelbrot set.

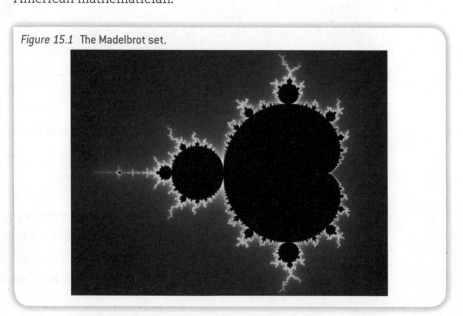

One of the aspects of fractals is "self-similarity". This means that the whole fractal is repeated in its parts, as can be seen in the picture of the Mandelbrot set on the previous page.

Examples of fractals in nature include trees, ferns, broccoli, coastlines and craters.

◇◇ WEB LINKS
To zoom in on parts of the Mandelbrot set, go to: www.youtube.com and search for "Deepest Mandelbrot Set Zoom".

 Activity 3   The Koch snowflake

The Koch snowflake was developed by Helge von Koch in the early 20th century. It is one of the earliest mathematical demonstrations of fractals.

To create the Koch snowflake, draw an equilateral triangle, such as the first picture shown in the diagram.

- Divide each side of the triangle into three equal lengths.
- Erase the middle piece of each side (the green lines shown).
- Use the erased side as the base of a new equilaterial triangle, as shown.
- Repeat this process again and again.

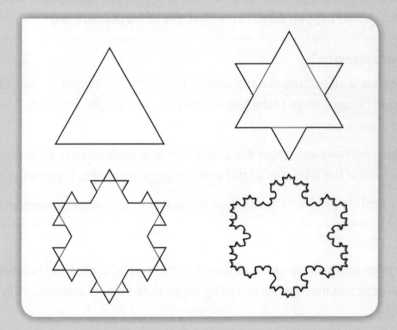

In the diagram, the process has been carried out three times. It shows three iterations.

Now try it for yourself. Investigate the perimeter and area of the resulting shape. Start with a large, equilateral triangle. Call the length of each side one unit. Let N = number of sides, L = length of one side, P = perimeter and A = area.

Copy the table on page 186 and fill in the information you know.

At the start, or stage 0, you have a triangle with three sides, each of length 1 unit, so the perimeter is 3 units and the area is $\frac{\sqrt{3}}{4}$ units^2.

Now go to the first stage, the second picture in the diagram, and continue to fill in the table. Do the same for the second and third stages.

Use simplified surd values for the table.

Iteration stage	N	L	P	A
0	3	1	3	$\frac{\sqrt{3}}{4}$
1	12			

a) Look at the emerging pattern in your table. What do you think the table values will be at the fourth stage? Verify your predictions by completing the process up to the fourth stage.

b) **Explain** the relationship of successive terms in each column of your table. Find a formula to identify the relationship mathematically.

c) Make four graphs of the individual four columns plotted against the stage value. Find a formula connecting the stage number to the column headings.

d) **Explain** what you think happens to the perimeter and area as the iteration, or stage number, increases. What would happen if you carried out this process endlessly?

RESEARCH   What is dimension?

All solid objects in our world have three dimensions: length, width and height. A point has no dimension, a line has one dimension (length) and a plane has two dimensions (length and width). Scientists consider time to be the fourth dimension.

A fractal, however, does not have an integer dimension. A fractal dimension is a number, or index, that expresses the complexity of the fractal as a ratio of the change in detail to the change in scale.

Research what the fractal dimension of the Koch snowflake is, and how the dimension is calculated.

REFLECTION

Geometrical objects often provide inspiration for artists. For example, Escher and Salvador Dali incorporated many mathematical ideas into their works, including impossible objects and objects in four dimensions, such as the hypercube.

Research an artist of your choice who has incorporated mathematics into their work. Write a short report about the aspects of mathematics explored and the artist's inspiration.

GLOBAL CONTEXTS
Personal and cultural expression

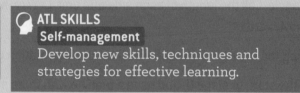
ATL SKILLS
Self-management
Develop new skills, techniques and strategies for effective learning.

TOPIC 3

Volume and surface area of 3D shapes

A huge amount of resources goes into the packaging of products that people buy every day. Consumers need to make wise choices about the products they buy. They should take note of the consequences of using non-renewable resources. Everyone wants packaging that will protect the product. If it becomes clear (through sales) that consumer preference is to minimize packaging, then suppliers will be more innovative and develop better packaging solutions that minimize wasted space and resources used.

To ensure sustainability through packaging, the health, safety and environmental effects of products must be monitored throughout their life cycle, beginning with consumer needs, through to the sourcing of raw materials and, finally, to the engineering and design of the product.

 Activity 4   Let's pack it in!

In an effort to be more environmentally conscious, one soup-manufacturing company is reducing the amount of cardboard they use in the packaging of cans as well as the space needed when shipping the cans. They are considering changing their strategy, to package 18 cans of soup together, instead of 12.

Traditionally the company used the **configuration** below.

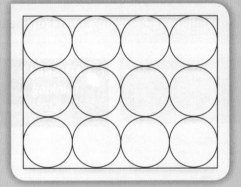

The 12 cylindrical cans are packed in a closed cardboard box shaped liked a rectangular prism.

Each can has a diameter of 9 cm and a height of 11 cm.

Determine the best possible way to configure 18 cans to minimize the cardboard used to make the packaging and minimize the wasted space in each box.

a) Suggest as many different configurations as possible. Include diagrams.

b) [Calculate] the dimensions of the rectangular box used in each configuration.

    i) [Calculate] the surface area required to make the box in each configuration.

    ii) To what degree of accuracy do you think your answers should be given?

c) Determine the optimal configuration to minimize the surface area and use the least amount of cardboard possible.

d) Given the original design with 12 cans, [calculate] the percentage increase or decrease in cardboard that will be used by moving to your 18-can configuration.

e) In order to reduce shipping costs, the other goal is to reduce the wasted space in each box. [Calculate] the wasted space in the box of the optimal configuration.

f) Given the original design with 12 cans, [calculate] the percentage increase or decrease in wasted space by moving to your 18-can configuration.

g) Based on your results, advise the company on whether they should move to an 18-can configuration.

Make sure that you include in your report:
- all calculations and diagrams
- a comparison of the different designs that you considered
- a description of which design you found optimal, with a labelled diagram
- a justification of the degree of accuracy used
- a comparison of the 12-can configuration to the 18-can configuration and your recommendation on which packaging strategy the company should use.

REFLECTION

    a) Think about the packaging that you see every day. Is most of it optimal?

    b) What factors do you think influence a company's decision on the shape of packaging to use?

    c) How do products with multiple layers of packaging impact on the use of resources?

GLOBAL CONTEXTS
Globalization and sustainability

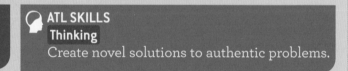
ATL SKILLS
Thinking
Create novel solutions to authentic problems.

Summary

Through the activities in this chapter, you have investigated how geometric properties of 2D and 3D space enable you to create objects such as mobiles and works of art. You applied what you have learned about the geometric properties of some objects to enable you to solve a real-world problem involving wasted space in packaging.

You have explored the mathematics behind impossible objects, or optical illusions. You have also seen that new geometries, such as fractal geometry, can extend understanding of irregular shapes and also redefine the concept of dimension.

16 System

A group of interrelated elements

INQUIRY QUESTIONS	**TOPIC 1** The real number system ■ **How do the properties of a system affect what you can accomplish within it?** **TOPIC 2** Geometric systems ■ **Can everything be defined?** **TOPIC 3** Probability systems ■ **How can a system save a life?**

COMMAND TERMS

Justify give valid reasons or evidence to support an answer or conclusion.

Prove use a sequence of logical steps to obtain the required result in a formal way.

Solve obtain the answer(s) using algebraic and/or numerical and/or graphical methods.

Use apply knowledge or rules to put theory into practice.

SKILLS

ATL

✓ Use and interpret a range of discipline-specific terms and symbols.

✓ Understand and use mathematical terminology.

✓ Understand and use mathematical notation.

✓ Make connections between subject groups and disciplines.

✓ Demonstrate persistence and perseverance.

Number

✓ Explain the components and properties of the real number system.

✓ Transform between different forms of number.

✓ Simplify numerical expressions.

Algebra

✓ Solve equations.

✓ Use set notation and terminology.

✓ Determine the intersection and union of sets.

✓ Expand and simplify algebraic expressions.

Geometry and trigonometry

✓ Explore the components and properties of a geometric system.

Statistics and probability

✓ Prove basic probability theorems.

✓ Apply the properties of probability systems to solve problems.

OTHER RELATED CONCEPTS

Equivalence   Generalization   Representation
Justification

Introducing system

What is a mathematical system? This is a very broad concept but, in general, a mathematical system is a collection of mathematical objects such as numbers or geometric objects, and a set of rules—postulates and **axioms**—that tell you what you are allowed do with these objects. Usually, a mathematical system also includes specific and rigorous definitions.

Pure mathematics is based on a system of axioms. An axiom is a statement that is accepted without proof. The properties of an axiomatic system are that it must be:

- consistent—it is free of any internal contradiction
- complete—every statement is derivable from the definitions and axioms.
- concise—it has internal simplicity and elegance.

Ever since you started school you have been studying, informally, different mathematical systems. For example, think of the natural numbers 0, 1, 2, 3, ... and the operation addition. The set of natural numbers under the operation of addition forms a mathematical system that can be described formally by a collection of statements called axioms.

Deduction is the type of reasoning that you use to build mathematical truths. Mathematicans use deductive reasoning to build mathematical proof. This means that each statement they make must follow logically from earlier statements. Those earlier statements could be the axioms, which they have agreed to be true at the start, or other statements that they have proved within this system. A proof in mathematics is an argument containing absolute certainty: it contains no doubt at all. Only axioms and theorems already proved can be used to prove new theorems.

What is an axiomatic system? It is a system with a starting position and a set of rules you are allowed to apply. Consider a game such as chess, or noughts and crosses (tic-tac-toe): you agree on a starting configuration and a set of rules for how the pieces may be moved. Whenever you play chess you move from one position to another by applying the rules. You might ask whether a certain position is valid within the game—this is the same thing as asking whether a mathematical result is true.

Despite the formal nature of an axiomatic system, most of the axioms are actually quite simple and intuitive. For example, the commutative axiom for the system of real numbers simply tells you that the order in which you add two real numbers doesn't matter, you will get the same result regardless of the order in which you add the numbers.

In this chapter, you will discover some mathematical systems that you have already been working with, and be invited to create one yourself.

TOPIC 1

The real number system

The real number system, together with the binary operations of addition and multiplication, consists of certain axioms, or properties, that you need to be familiar with before doing this activity.

 Activity 1   **No shortcuts allowed!**

STEP 1 **a)** Research the following properties of the real number system with the binary operations of addition and multiplication. Write down the properties, including an example of each.
- Closure property
- Commutative property
- Associative property
- Identity property
- Inverse property
- Distributive property for addition over multiplication

b) Do these properties hold for the operations of subtraction and division in the set of real numbers?

c) What are the addition, multiplication and substitution principles for solving equations?

d) Use these properties and principles to justify your solution of an equation. Study the following example first. The left-hand column shows the equation and the changes it goes through as you apply certain properties. The right-hand column lists the justifications for each of those changes.

Solve for a:

Equation	Reasons
$2a + 3 = 9$	Given
$(2a + 3) + {-3} = 9 + {-3}$	Addition principle
$2a + (3 + {-3}) = 6$	Associative property of addition
$2a + 0 = 6$	Inverse property of addition
$2a = 6$	Identity property of addition
$\dfrac{1}{2} \times (2a) = \dfrac{1}{2}(6)$	Multiplication principle

$$\left(\frac{1}{2}\times2\right)a=3 \qquad \text{Associative property of multiplication}$$

$$1a=3 \qquad \text{Inverse property of multiplication}$$

$$a=3 \qquad \text{Identity property of multiplication}$$

Did you know that you go through all the stages listed above, when you use the shortcuts you have learned to solve this simple equation? The properties above are the axioms, or "behind the scenes" accepted reasons, that allow you to solve the equation in just two or three lines.

STEP 2 **a)** **Solve** these equations. Use the above properties to **justify** each stage.

    i) $4+3x=19$

    ii) $5-2x=6$

    iii) $2x+3x=10$

    iv) $\dfrac{2x}{3}-5=3$

b) Suggest why pure mathematicians might claim that subtraction and division do not really exist as operations. Think of what these two operations mean in terms of addition and multiplication.

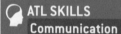

ATL SKILLS

Communication

Use and interpret a range of discipline-specific terms and symbols.

Activity 2 **Testing the axioms**

In this activity you will use the properties: commutative, associative, identity and inverse.

  a) Determine which of the properties listed in Activity 1, if any, are satisfied for each of these sets under the given operation.

    i) The set of real numbers under the operation of (1) subtraction, and (2) division.

    ii) The set of natural numbers under the operation of (1) addition, and (2) multiplication.

Not all binary operations in a given set satisfy all the properties listed above. For example, define $a \ast b$ to be a^b. Then, if a and b are 2 and 3 respectively, $2 \ast 3 = 2^3 = 8$, and $3 \ast 2 = 3^2 = 9$. Hence, in the set of natural numbers, $a \ast b$ is not commutative since $2^3 \neq 3^2$. Is $\ast$ associative? Is $(a \ast b) \ast c = a \ast (b \ast c)$, for all natural numbers a, b and c?

Using 2, 3 and 4 for a, b and c respectively, $(2 \ast 3) \ast 4 = (2^3)^4 = 2^{12}$. $2 \ast (3 \ast 4) = 2^{3 \ast 4} = 2^{3^4} = 2^{81}$. Since $2^{12} \neq 2^{81}$, $\ast$ is not associative.

TIP

Remember, to show that a certain property is not true, it is sufficient to find one counter-example where the property does not hold.

b) For each of the binary operations on the given set of numbers listed below, determine whether the commutative and associative properties hold. If it does not hold, show a counter-example.

   i) The binary operation "the mean of" under the set of natural numbers, that is, for any two natural numbers a and b, $a \# b = \dfrac{a+b}{2}$.

   ii) The binary operations of "intersection" and "union" for any two sets A and B, that is, (1) $A \cap B$ and (2) $A \cup B$.

   iii) Define the binary operation $\star$ as $a \star b = 2^{a+b}$ under the set of real numbers.

   iv) Define the binary operation # as $a \# b = a - a^{b+b}$ under the set of real numbers.

c) Create two different binary operations on a particular set. Determine if they satisfy the commutative and associative properties.

ATL SKILLS

Communication

Use and interpret a range of discipline-specific terms and symbols.

 **Activity 3**   **Is being "irrational" like being "complex"?**

The real number system consists of the set of real numbers and the properties that govern this set under a given binary operation. The set of real numbers, designated by $\mathbb{R}$, consists of the following subsets of numbers: integers, designated by $\mathbb{Z}$, natural numbers, designated by $\mathbb{N}$, rational numbers, designated by $\mathbb{Q}$, and irrational numbers, that is, any number that is not rational. There are other subsets that are defined according to their sign, for example, $\mathbb{Z}^+$, the set of positive integers.

In this activity you will consider two types of numbers, one which is a real number, and one which is not real, but nonetheless they enjoy some similarities.

STEP 1   **Rational versus irrational**

   a) A rational number is any real number that can be expressed in the form $\dfrac{p}{q}$ where p and q are integers but $q \neq 0$. Using set notation, $\mathbb{Q} = \left\{ \dfrac{p}{q} \mid p, q \in \mathbb{Z}, q \neq 0 \right\}$. The vertical slash stands for "given that". Show that each of these numbers is rational.

   i) 0.5   **ii)** −3   **iii)** 0.222…   **iv)** 0.8333…   **v)** 1.245

   b) What types of decimal number are rational? Explain.

   c) An irrational number is any real number that is not rational, that is, it cannot be expressed in the form $\dfrac{p}{q}$ where p and q are integers and $q \neq 0$. A common example is the number π. Explain why π is irrational.

   d) Research other famous irrational numbers and explain how you know they are irrational. Give examples of where you have worked with these numbers.

Operations with irrational numbers

In this task, consider only irrational numbers of the form $a\sqrt{b}$.

a) Perform the following operations.

  i) $2\sqrt{3} + 4\sqrt{3}$

  iii) $4\sqrt{7} + \sqrt{7}$

  v) $(-\sqrt{3} + 6\sqrt{2})(3\sqrt{3} - \sqrt{2})$

  vii) $(-5\sqrt{5} - 2\sqrt{6})(3\sqrt{5} - 4\sqrt{6})$

  ix) $(3 - 4\sqrt{7})(3 + 4\sqrt{7})$

  ii) $3\sqrt{5} - 7\sqrt{5}$

  iv) $(2 + 4\sqrt{2})(3 - \sqrt{2})$

  vi) $(\sqrt{7} + 2\sqrt{2})(\sqrt{7} - 2\sqrt{2})$

  viii) $(2\sqrt{2} + \sqrt{3})^2$

b) State the axioms you used as you performed these operations. State the axioms that you used.

c) Although you started with irrational numbers, some of your answers were rational numbers. Explain why this happens.

Calculating with irrational numbers can be very similar to using the same operations with polynomials. When you need to divide a number by an irrational number that contains a radical, start by multiplying both terms by the conjugate of the denominator. This is called rationalization of the denominator. Based on your answer to question **c)**, how would you rationalize the denominator for this fraction?

$$\frac{2+\sqrt{3}}{1-\sqrt{3}}$$

TIP

Use the identity axiom for multiplication.

d) Rationalize the denominator in each of these fractions.

  i) $\dfrac{1+2\sqrt{3}}{4-\sqrt{3}}$

  ii) $\dfrac{2-3\sqrt{2}}{3+\sqrt{2}}$

  iii) $\dfrac{1-4\sqrt{5}}{2-3\sqrt{5}}$

  iv) $\dfrac{5+6\sqrt{2}}{7+3\sqrt{3}}$

e) Why is "rationalizing the denominator" an appropriate name?

f) Is it possible to rationalize the numerator and the denominator of a fraction? Explain.

Complex numbers

a) Research complex numbers and their history. What are they used for? Why are they necessary?

Complex numbers take the form $z = a + bi$, where a and b are real numbers, and $i = \sqrt{-1}$. Using set notation, $\mathbb{C} = \{a + bi \mid a, b \in \mathbb{R}, i = \sqrt{-1}\}$. a is the real part and b is the imaginary part of the number since i is an imaginary number (hence, it's not a real number). The operations addition and multiplication of complex numbers have exactly the same properties as addition and multiplication of real numbers.

b) Complete each of these operations (remembering that $i^2 = -1$).

  i) $2i + 4i$

  iv) $(2 + 4i)(3 - i)$

  vii) $(-5 - 2i)(3 - 4i)$

  ii) $3i - 7i$

  v) $(-3 + 6i)(3 - i)$

  viii) $(2 + i)^2$

  iii) $4i + i$

  vi) $(7 + 2i)(7 - 2i)$

  ix) $(3 - 4i)(3 + 4i)$

c) Based on what you have learned so far, how can you eliminate the complex number in the denominator of this fraction?

$$\frac{2+i}{1-i}$$

d) Simplify these fractions in the same way.

   i) $\dfrac{1+2i}{4-i}$ **ii)** $\dfrac{2-3i}{3+i}$ **iii)** $\dfrac{1-4i}{2-3i}$ **iv)** $\dfrac{5+6i}{7+3i}$

e) Why is it possible to divide complex numbers in the same way as irrational numbers?

REFLECTION

a) Are real numbers also complex numbers? Are complex numbers also real numbers? Explain your answers.

b) Draw a suitable diagram, for example, a Venn diagram to illustrate the relationship between the following subsets of real numbers:

natural numbers, integers, rational numbers, irrational numbers.

Do the complex numbers have a place in your diagram? Explain.

c) Determine if the set of complex numbers with the binary operations of addition and multiplication have the same properties as real numbers under the same binary operations.

d) Research how to use the Argand diagram to graph complex numbers. Show some examples.

ATL SKILLS
Communication
Use and interpret a range of discipline-specific terms and symbols.

TOPIC 2

Geometric systems

A finite geometrical system is unlike the geometry you study in school in that it uses only a limited number of objects. Creating such a system helps you to understand infinite axiomatic systems, such as Euclidean geometry. This is the geometry with which you are familiar.

The geometry you study in school is the same geometry that was developed by Euclid of Alexandria in 300 BC in his famous book *The Elements*. The book is a collection of logical conclusions deduced from axioms or postulates, or truths that are self-evident. Euclidean geometry is an infinite geometric system. There are other infinite non-Euclidean geometric systems that were mainly developed by two mathematicians, Riemann and Lobachevsky, in the late 18th and early 19th centuries.

🏛 **MATHS THROUGH HISTORY**
Born in the 20th century, the Polish mathematician Alfred Tarski helped to modernize the geometry you study in school. He did this by using logic and philosophy. Throughout his lifetime, he was concerned with the problem of mathematical truth. He is well known for the proof of the ball paradox. This states that if a ball is cut into an infinite number of pieces, it can be reassembled into two balls, the sizes of which will be equal to the size of the original ball!

The study of finite geometrical systems is a relatively new branch in mathematics. Its development is mainly attributable to the Italian mathematician Gino Fano, who worked on it in the early 20th century. Since its recent birth, this field has found many real-world applications, especially in the area of statistical designs. One of these, called *block design*, is being used to test new medical drugs on patients. In the next activity you will explore the basic aspects of Fano's finite geometry.

 Activity 4 **Creating an axiomatic system**

This system will have two undefined terms: *point* and *line*.

Here are the axioms to go with these undefined terms.

A1: There exists at least one line.

A2: Every line has exactly three points.

A3: Not all points are on the same line.

A4: For any two points, there exists exactly one line that passes through both of them.

A5: Any two different lines have at least one point in common.

a) How do the geometry you have studied compare with the system above. How many undefined terms does Euclidean geometry have? Name the term(s).

b) How do the axioms above differ from the axioms for points and lines that you know?

c) Deduce these properties from the finite geometry system.

   i) Consider only axioms 1 and 2 above. What is the minimum number of points in this geometry?

   ii) Consider only axioms 1 to 3 above. What is the minimum number of points in this geometry?

   iii) Consider all five axioms. Determine the number of points and lines in this geometry.

   iv) Use the number of points and lines from part iii) to create a model of this geometry.

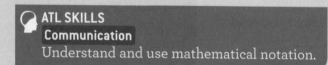

ATL SKILLS
Communication
Understand and use mathematical notation.

Probability systems

A simple example of a mathematical system is a system of axioms for probability theory: it consists of just three axioms, $A1$, $A2$ and $A3$, and two basic definitions, $D1$ and $D2$.

Let $S = \{a_1, ..., a_n\}$ be a sample space. This means the number of elements in S is n.

Let $P(A)$ be the probability of outcome A happening, and $P(S)$ be the probability of outcome S happening.

A1: $P(A) \geqslant 0$, for any outcome $A \subseteq S$

A2: $P(S) = 1$

A3: If $A \cap B = \varnothing$ then $P(A \cup B) = P(A) + P(B)$

D1: An outcome A is any subset of S. If $A = \varnothing$ then A is the impossible outcome; if $A = \{a_k\}$ for some $k \in \{1, ..., n\}$ then A is a single outcome.

For example, suppose you roll a die.
The sample space is $S = \{1, 2, 3, 4, 5, 6\}$ and the single outcomes are the subsets each consisting of one of these numbers.

D2: Given an outcome A, the complement of A is the outcome A' that verifies the following properties:

    (i) $A \cup A' = S$ **(ii)** $A \cap A' = \varnothing$

For example, when rolling a die $\{1, 2, 3\}$ and $\{4, 5, 6\}$ are complementary outcomes.

You can use this system of axioms to deduce some simple probability theorems. Although they are obvious, you do not need to accept them as axioms since they can be deduced from $A1$–$A3$. Remember that a system of axioms should be kept concise, that is, include the minimum number of statements that allow you to prove the rest of the results in the theory under study.

For example, the first theorem, $T1$, is a rather obvious statement:

T1: $P(A) \leqslant 1$, for any event $A \subseteq S$

Proof: Given any event A, *by definition*, $A \cup A' = S$, since by **A2**, $P(S) = 1$, and you can conclude that $P(A \cup A') = 1$.

By **A3**, as $A \cap A' = \varnothing$, $P(A \cup A') = P(A) + P(A')$

So, $1 = P(A) + P(A')$. (Why?)

By **A1**, $P(A) \geqslant 0$ and $P(A') \geqslant 0$.

So, $P(A) \leqslant 1$.

Activity 5   Using axioms

Use the axioms A1–A3 and any theorem already proved, to deduce the following results.

T2: $P(A') = 1 - P(A)$, for any outcome $A \subseteq S$.

T3: If $S = \{a_1, ..., a_n\}$ is a sample space where all the single outcomes have the same probability, then for any outcome A, so that $n(A) = k$, $P(A) = \dfrac{k}{n}$.

T4: $P(A \cup B) = P(A) + P(B) - P(A \cap B)$, for any outcomes A and B.

T5: $P(\emptyset) = 0$

> **TIP**
>
> Venn diagrams are an effective tool in solving probability problems.

> **ATL SKILLS**
> **Communication**
> Understand and use mathematical notation.

> 🏛 **MATHS THROUGH HISTORY**
> In 1930, the Nobel Prize in Physiology and Medicine was awarded to the Austrian biologist and physician Karl Landsteiner for his discovery of human blood groups. In 1940, he was also part of the group that discovered the rhesus factor component of blood. After his successful experimental work, blood transfusions became safer and fewer people died as a result of transfusions.

Blood typing

Human blood is living tissue composed of cells suspended in plasma, which is the yellow fluid component of blood. Red blood cells (carry oxygen), white blood cells (fight infection) and platelets (stop bleeding in injuries) make up about 45 per cent of blood. The rest is plasma. Blood accounts for about 7 per cent of your body weight. An average-sized adult female's body contains about 4.5 litres (9 pints) of blood. An average-sized adult male contains about 5.5 litres (12 pints) of blood.

Human blood can be classified into four major blood types: O, A, B and AB. There are also subtypes of these, for example, A1, A2. The blood type also has an associated antigen called the rhesus factor. An antigen is a protein or carbohydrate substance that the body recognizes as being foreign. About 85 per cent of all people possess this antigen and their rhesus factor (Rh) is therefore positive. The rest, about 15 per cent, are Rh negative. This antigen was first found in rhesus monkeys, hence its name.

If a person begins to lose blood through an accident or surgical procedure, for example, it is important to try to control the loss of blood. If it is not possible to control the blood loss, then a blood transfusion will be necessary, to save the person's life. In the United States alone, about every two seconds someone needs a blood transfusion. Blood typing is therefore essential so that blood can be donated and received.

When a person needs a transfusion, it is best to give them blood that matches their own blood type and rhesus factor. This is not always possible, so the second best choice is a blood type that is compatible with that of the patient. Look at the following blood compatibility chart.

You can receive type:								
If you are type:	0+	0−	A+	A−	B+	B−	AB+	AB−
0+	*	*						
0−		*						
A+	*	*	*	*				
A−		*		*				
B+	*	*			*	*		
B−		*				*		
AB+	*	*	*	*	*	*	*	*
AB−		*		*		*		*

Note that, for example, blood type A+ can only donate to blood types A and AB, but can receive from A+, A−, O+ and O−.

Administering a blood transfusion of an incompatible blood type leads to toxic reactions in the patient. These usually lead to death. It is essential to administer the correct blood type to the patient.

👤 Activity 6 Blood typing

STEP 1 Study the blood compatibility chart and the example shown below for A+. From the blood type compatibility chart, devise a system to complete the chart below as efficiently as possible.

Blood type	Can donate blood to	Can receive blood from
0+		
0−		
A+	A+, AB+	A+, A−, 0+, 0−
A−		
B+		
B−		
AB+		
AB−		

STEP 2 **a)** One blood type is considered to be the universal donor. Which one? Why?

b) One blood type is considered to be the universal recipient. Which one? Why?

Before you continue, your teacher will show you how to play the blood typing game found at: www.nobelprize.org (go to *Educational > Play the Blood Typing Game*). After successfully saving your patients by administering the correct blood transfusions, continue with activity 7.

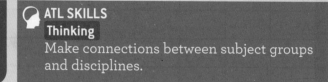

GLOBAL CONTEXTS
Scientific and technical innovation

ATL SKILLS
Thinking
Make connections between subject groups and disciplines.

Activity 7   Blood typing 2

a) Use the information from your table in Activity 6 to fill in columns 2 and 3 of the table below. The figures in column 4 of the table give the percentages of people in the United States with the particular blood type. The figures are taken from statistics at the blood center of Stanford University. These percentages are approximate and will vary according to gender and race.

Use these percentages to calculate the probability of finding a compatible donor for each blood type. Enter these figures in column 5. The probabilities for A+ are done for you. The probability of finding a compatible donor for blood type A+ is the sum of the probabilities of people in the general population that have A+, A−, O+ and O−, or $35.7 + 6.3 + 37.4 + 6.6$, or 86 per cent.

Blood type	Can donate blood to	Can receive blood from	% of general population	Probability of finding a compatible donor
O+			37.4	
O−			6.6	
A+	A+, AB+	A+, A−, O+, O−	35.7	86%
A−			6.3	
B+			8.5	
B−			1.5	
AB+			3.4	
AB−			0.6	

b) State the probability axiom that you are using to calculate the figures in column 5 of the chart. Justify the use of this axiom in this situation.

c) Rank the blood types in order of probability of finding a compatible donor. List them from highest to lowest probability.

d) Suppose a person in the USA is selected randomly. Let A be the outcome that the person has blood type A, let B be the outcome that the person has blood type B, and so on. **Use** the percentages in your chart to find the following probabilities.

    i) $P(A \text{ or } B)$

    ii) $P((A'))$

    iii) $P((AB)')$

    iv) $P(AB \text{ or } O)$

e) State in which cases the probabilities in **d** correspond to probabilities of complementary events.

f) Determine the percentage of people who are Rh+ and Rh−, then find the probabilities for the following non-mutually exclusive events.

    i) $P(B \text{ or Rh}-)$

    ii) $P(AB \text{ or Rh}+)$

    iii) $P(A \text{ or } B \mid \text{Rh}+)$ This is read: the probability that a randomly selected person has blood type A or B given that the person is Rh+. The slash stands for "given that".

    iv) $P(O \text{ or } AB \mid \text{Rh}-)$

    v) $P(\text{Rh}+ \mid AB \text{ or } B)$

    vi) $P(\text{Rh}- \mid O \text{ or } A)$

g) The statistics for blood types vary from country to country. In the chart below, the blood type statistics for Australia and Finland are combined, so that the total for each column represents 50 per cent. **Use** the chart to answer the questions that follow. Explain how your answers show that blood type and origin of country are not independent.

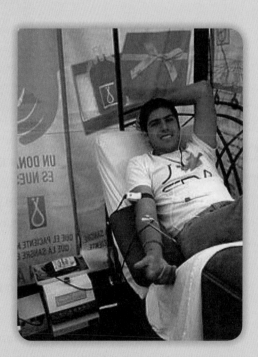

Blood Type	Probabilities		
	Australia	Finland	Total
0	0.245	0.155	0.40
A	0.19	0.22	0.41
B	0.05	0.085	0.135
AB	0.015	0.04	0.055
Total	0.50	0.50	1.0

    i) $P(A \mid \text{Finland})$ **iii)** $P(AB \mid \text{Finland})$

    ii) $P(A \mid \text{Australia})$ **iv)** $P(AB \mid \text{Australia})$

REFLECTION

a) Suppose you were organizing a blood donor campaign for your local hospital's blood bank. Based on the statistics you calculated in this topic, what blood types would you focus on?

b) According to blood bank statistics, people with O+ and A+ tend not to donate. When surveyed, potential donors said that their blood is not needed because they are of a common blood type. Is this a sensible reason? Explain why or why not.

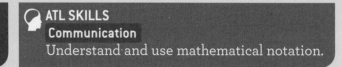

🌐 **GLOBAL CONTEXTS**
Scientific and technical innovation

🧠 **ATL SKILLS**
Communication
Understand and use mathematical notation.

Exploring properties of axiomatic systems

The properties of an axiomatic system can be summarized as the 3Cs: complete, concise and consistent. You will use these properties to solve the next problem.

🏛 **MATHS THROUGH HISTORY**

David Hilbert, a German mathematician, was probably the most influential mathematician of the 19th and 20th centuries. It was Hilbert who exerted the greatest influence on the development of Euclidean geometry, after Euclid himself. His work, entitled *Grundlagen der Geometrie* (*The Foundations of Geometry*), 1899, was considered so important that nine editions came out during the 60 years of its publication. Since his first attempt, the axiomatic approach has remained a permanent feature of mathematics.

Activity 8   Bringing a ship to dock

This problem is based on a series of word puzzles written by Lewis Carroll, the author of *Alice in Wonderland*, in the late 19th century. Using the rules below, you start with the word SHIP and end up with the word DOCK.

- Each word must be a real four-letter English word found in any standard English dictionary.
- Each word is identical to the previous word, except for one letter.
- No rude words are allowed!

While you are solving this puzzle, be aware of the strategies you are using. In particular, think about the relationship of solving this puzzle to proving a theorem by using deduction.

You may use the following axiom in solving the puzzle.

Axiom: All proper words in the English language contain at least one vowel.

One possible way to begin is to change the H to L.

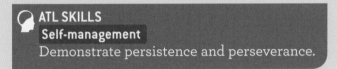

S H I P
S L I P

Use this first stage, or choose one of your own. Follow the rules and the axiom to find a possible solution. Then, see if you can improve upon your solution. Remember that one of the characteristics of an axiomatic system is that it is concise, so see if you can bring the ship to dock in the fewest possible moves!

After you have found your shortest solution, answer this question.

- How many possible solutions do you think this puzzle has, given that words may be repeated?

Look at your solutions and make a conjecture regarding the number of vowels in your words. **Prove** your conjecture.

ATL SKILLS
Self-management
Demonstrate persistence and perseverance.

Summary

In this chapter you have seen some practical applications of mathematical systems, and as you worked through them, you almost certainly realized that "abstract" does not necessarily mean "difficult".

Understanding and working with mathematical systems should improve your ability to think logically, since they necessitate working in an organized and systematic way. They should provide the opportunity to evaluate your own methods of reasoning when solving problems:

- Is every step well thought out and justified?
- Does every step follow on logically from the previous one?
- Have you used the minimum amount of steps to achieve your goal?

Notes

Notes

Notes

Notes

Notes

Notes

Notes

Notes

Notes

Notes

Notes

Notes

Notes